98-REM-1a

APPLICATION OF A DYNAMIC MODEL TO ASSESS MICROBIAL HEALTH RISKS ASSOCIATED WITH BENEFICIAL USES OF BIOSOLIDS

by:

Joseph N.S. Eisenberg, Ph.D.
School of Public Health, University of California, Berkeley

Kelly Moore
John M. Colford, Jr., MD, Ph.D.
School of Public Health, University of California, Berkeley

Jeffrey A. Soller
Don M. Eisenberg, Ph.D.
Eisenberg, Olivieri and Associates, Inc.

2006

WERF

The Water Environment Research Foundation, a not-for-profit organization, funds and manages water quality research for its subscribers through a diverse public-private partnership between municipal utilities, corporations, academia, industry, and the federal government. WERF subscribers include municipal and regional water and wastewater utilities, industrial corporations, environmental engineering firms, and others that share a commitment to cost-effective water quality solutions. WERF is dedicated to advancing science and technology addressing water quality issues as they impact water resources, the atmosphere, the lands, and quality of life.

For more information, contact:
Water Environment Research Foundation
635 Slaters Lane, Suite 300
Alexandria, VA 22314-1177
Tel: (703) 684-2470
Fax: (703) 299-0742
www.werf.org
werf@werf.org

This report was co-published by the following organization. For non-subscriber sales information, contact:

IWA Publishing
Alliance House, 12 Caxton Street
London SW1H 0QS, United Kingdom
Tel: +44 (0) 20 7654 5500
Fax: +44 (0) 20 7654 5555
www.iwapublishing.com
publications@iwap.co.uk

Library of Congress Catalog Card Number: 2006924261
IWAP ISBN: 978-1-84339-762-5

School of Public Health, University of California, Berkeley

The research on which this report is based was funded, in part, by the United States Environmental Protection Agency through Cooperative Agreement No. CR-827345-01 with the Water Environment Research Foundation (WERF). Unless an EPA logo appears on the cover, this report is a publication of WERF, not EPA. Funds awarded under the Cooperative Agreement cited above were not used for editorial services, reproduction, printing, or distribution.

ABSTRACT AND BENEFITS

Abstract:

This investigation involved the development and demonstration of a risk-based methodology for assessing the risk to human health from exposure to pathogens via biosolids. Currently, pollutant limits for chemical constituents in biosolids are based on the results of quantitative risk assessments, however pathogens in biosolids are not based on risk assessments. In this report three sources of environmental data were used to estimate risk: pathogen monitoring of raw sludge, efficacy of treatment processes, and pathogen monitoring of post-treatment biosolids. Risks were estimated, even when post-treatment monitoring data are all below the detectable level, for a number of different models to gain insight into risks associated with different aspects of the biosolids treatment and application process. Biosolids producers can use this risk-based approach to interpret the results of biosolids monitoring data with respect to health implications. Treatment plant engineers can evaluate the risk-based benefits of making operational changes to existing treatment and/or adding additional treatment processes. Environmental managers can evaluate potential capital improvements and/or land application site placement issues from a health-based perspective. Finally, the regulation of pathogens can now be based on human health risk in a manner parallel to that which is used for chemical contaminants.

Benefits:

- Provides a structure and methodology from which risk-based regulations can be developed for biosolids-related exposures to pathogens.
- Provides a means to estimate the number of infections and illnesses in a population that may be associated with biosolids exposure.
- Provides a structure to estimate risks that is consistent with the current understanding of infectious disease processes.
- Accounts for uncertainty in risk estimation so that decision making can be more informed.
- The model allows for the incorporation of future data related to biosolids and/or pathogens.

Keywords: Biosolids, microbial risk assessment, pathogens

ACKNOWLEDGMENTS

The authors thank the members of the project's health Advisory Committee: Phillip Berger, Walter Jakubowski, Greg Kester, Suresh Pillai, and Robert Spear. Current affilation for J.N.S. Eisenberg is the School of Public Health, University of Michigan, Ann Arbor.

Report Preparation

Principal Investigator:
Joseph N.S. Eisenberg, Ph.D.
School of Public Health, University of California, Berkeley

Project Team:
Kelly Moore
John M. Colford, Jr., M.D., Ph.D.
School of Public Health, University of California, Berkeley
Jeffrey A. Soller
Don M. Eisenberg, Ph.D.
Eisenberg, Olivieri and Associates, Inc.

Project Subcommittee

Cecil Lue-Hing, D.Sc., P.E., D.E.E.
Cecil Lue-Hing & Associates, Inc.
Sydney Munger
Parametrix, Inc.
Rick Danielson, Ph.D.
Biovir, Inc.
Thomas Murphy, P.E.
New York City Dept. of Environmental Protection
Salvador Sedita, Ph.D.
Consultant
James E. Smith, Jr., D.Sc.
U.S. Environmental Protection Agency - NRMRL
George Tchobanoglous, Ph.D.
Tchobanoglous Consulting

Water Environment Research Foundation Staff

Director of Research: Daniel M. Woltering, Ph.D.
Program Manager: Lola Olaboda, M.P.H.

TABLE OF CONTENTS

LIST OF TABLES

LIST OF FIGURES

LIST OF ACRONYMS

AR	Attributable Risk
CI	Cumulative Incidence
g	gram
h	hour
HAC	Health Advisory Committee
HEI	Highly Exposed Individual
ILSI	International Life Sciences Institute
IU	Infectious Units
mg/day	milligram per day
mL	milliliter
MPN	Most Probable Number
NRC	National Research Council
PFU	plaque forming units
PSC	Project Subcommittee
RT-PCR	Reverse Transcriptase Polymerase Chain Reaction
TS	Total Solids
U.S. EPA	United States Environmental Protection Agency
WERF	Water Environment Research Foundation

EXECUTIVE SUMMARY

ES.1 Motivation for a Risk-Based Approach to Regulation and Decision Making

Treated sewage sludge, referred to as biosolids, has been used as a soil amendment for agricultural purposes since the onset of municipal wastewater treatment, and its use has intensified over the past 20 years. In 1993, the Environmental Protection Agency (U.S. EPA) promulgated the Part 503 Rule that set standards for use and disposal of sewage sludge.

Part 503 established pathogen control requirements based on technological and operational standards rather than on direct risk assessment. In recent years there has been an increasing polarization among advocates and critics of the use of biosolids, which has raised questions about the adequacy of the pathogen standards for protecting public health. In response to this polarization the U.S. EPA requested the formation of a National Research Council (NRC) committee to review the adequacy of the regulations including the technical basis of the pathogen requirements for biosolids, and the feasibility of utilizing microbial risk assessment. The conclusions reached by the committee were that, "there is no documented scientific evidence that the Part 503 rule has failed to protect public health. However, additional scientific work is needed to reduce persistent uncertainty about the potential for adverse human health effects from exposure to biosolids." Epidemiology provides one approach to reducing this uncertainty by directly estimating health effects. Two sources of data can be used for obtaining direct estimates of health risks: 1) passive surveillance, which relies on the reporting of cases to health agencies; and 2) observational studies, which actively collect exposure and outcome data on individuals. Passive surveillance can generally only detect high-risk conditions that result in larger outbreaks. Observational studies are needed to improve the sensitivity of detecting risks. However, even the improved sensitivity obtained through the use of observational studies in estimating risk generally does not provide low enough sensitivity to be acceptable by the U.S. EPA for developing regulations. Therefore, the absence of health data showing an association between illness and biosolids exposure does not mean that regulation is unwarranted.

Due to this lack of epidemiologically-based direct measures of risk to environmental exposures, the U.S. EPA and others have employed risk assessment to provide a basis for regulation and make operational decisions. This was essentially echoed in the NRC report that recommended the use of "improved risk-assessment methods to better establish standards for chemicals and pathogens." Risk assessment methods estimate the risk given environmental data on pathogen occurrence. A risk assessment provides a structured, scientifically defensible approach to estimating risks, even when risks are below the detection limit of epidemiology and exposure measurements.

Since the promulgation of the Final Part 503 Rule for pathogens was not based on risk assessment, the U.S. EPA has been committed to the reformulation of this regulation using a risk-based approach. This report presents a comprehensive method for characterizing risk due to exposure to microbial pathogens in biosolids.

ES.2 Overview of Risk-based Approach

The approach detailed in this report couples three data sources, pathogen data from raw sludge, treatment process efficacy data, and post treatment pathogen monitoring data with a health-based risk model. The use of all three data sources provides a means to estimate risk even when post-treatment pathogen data are all below the detectable level. The incorporation of raw sludge data also provides a means to examine the effects of variation in influent concentration in the risk estimate. The explicit modeling of the treatment process allows for the evaluation of the health impact due to environmental changes in the context of an existing treatment process or due to design improvements to the treatment process. Even if the post treatment monitoring data are below detectable levels, the modeling ensures that risk estimates are consistent with these nondetectable pathogen data.

Since are no federal requirements to monitor pathogens in raw sludge or even in most finished Class A biosolids or any Class B biosolids, there is a paucity of such data available. Even though analytical data are limited, the methodology herein provides a reliable model for performing risk assessment. As more data become available, the model outputs will likewise become more robust. The method presented herein provides a means to obtain risk estimates for Class A and Class B biosolids, even when having detectable post-treatment pathogen data are not required.

ES.3 Methods

The risk assessment model is a two-step process. First an exposure model estimates a dose that humans are exposed to, and secondly, a health effects model estimates risk. The exposure model consists of four components:

1. **Modeling pathogen concentrations in raw sludge**. Data are fit to a lognormal distribution, using a regression on order statistics (ROS) technique to impute nondetectable values.
2. **Modeling the treatment process**. The model used in this report comprises a combination of either one or two mesophilic anaerobic digesters with or without lime treatment.
3. **Modeling pathogen concentrations in biosolids (post-treatment)**. These data are used to constrain risk estimates; i.e., predicted pathogen concentrations from the treatment process model must be consistent with these monitoring data.
4. **Modeling exposure pathways**. The pathways examined include direct ingestion, groundwater, and aerosol. Direct ingestion focuses both on children living in nearby residences and occupational workers in the biosolids treatment and/or application.

The health effects model takes the dose estimated from the exposure model and estimates risk. Risk, in this report is presented in three ways: 1) single-event risks, which provide information on the probability of infection or disease of an individual who is exposed to biosolids as a single event; 2) annual risk estimates, which take into account the amount of biosolids exposure during a given year as well as the target population that is exposed. This estimates the probability of infection or sickness for an individual exposed to biosolids over an entire year; and 3) attributable risk estimates, which take into account secondary transmission and is defined as the number of cases of infection or disease that is attributable to exposure to biosolids and differentiating those cases from all other causes of infection or disease.

ES.4 Findings

Significant findings include:

1. The risk assessment approach developed here is able to estimate risks even for Class A biosolids where post-treatment monitoring data are all below the detectable level.
2. Further data collection should be expanded to include: Class A and B monitoring of pathogens both before and after treatment; and improved characterization of treatment processes with an emphasis on variability due to environmental and operational conditions.
3. The uncertainty and variability in pathogen concentrations influences risk estimates. Improved sampling can decrease uncertainty and improve the precision of risk estimates. Treatment process design can help decrease variability, which in turn can decrease mean risk estimates. For example, our analysis suggests that adding a second anaerobic digester will significantly decrease the probability risk.
4. Our groundwater and aerosol exposure pathway analysis can be used by regulators to develop risk-based set back distances from residences and from karst formations.
5. Population-based models provide a means to explicitly account for secondary transmission and immunity.

ES.5 Conclusions

This project demonstrated that comprehensive risk assessments for biosolids exposure are practical. Risk estimates are provided to illustrate how the risk modeling approach presented in this report can be used both in a regulatory context to develop risk-based rules, and in an operational context to examine the benefits of changing treatment processes in the context of current application practices or proposed new practices. The risk assessment framework presented in this report provides a mechanism to discuss the microbial risk associated with biosolids management using a common metric for comparison of treatment methods, management alternatives, and potentially to set risk-based standards for microbial contaminants in biosolids.

CHAPTER 1.0

INTRODUCTION

1.1 Motivation for a Risk-based Approach to Regulation and Decision Making

In 1993, the U.S. EPA published in the Federal Register the Part 503 rule that protects public health and the environment from adverse effects of pollutants that may be present in biosolids (U.S. EPA, 1995). In the Part 503 rule, pollutant limits for chemical constituents in biosolids were based on the results of risk assessments that were conducted to identify what, if any risks were associated with the use or disposal of biosolids. Pollutant limits for pathogens, however, were not based on risk assessments; rather they were based on either performance, technology-based standards, or on management and record keeping practices shown to protect human health and the environment. At the time, the U.S. EPA thought that methodologies had not been developed sufficiently to make risk-based calculations, especially given the lack of exposure data (U.S. EPA, 1989b, U.S. EPA, 1992b, U.S. EPA, 1995). This report presents an approach to conduct risk assessments associated with biosolids exposures, given the improved but still minimal availability of exposure data. The model as demonstrated herein provides the necessary framework for microbial risk assessment, both providing risk estimates and identifying data gaps that when filled will provide more robust estimates.

The U.S. EPA has a long history of using risk-based approaches for regulatory purposes, specifically to address their mandate to protect public health from environmental exposures to waterborne enteric pathogens. The use of risk assessment to address concerns with water related risks have partially been due to well-documented outbreaks associated with drinking water and recreational water exposures. Additional health concern has been due to studies of risks in nonoutbreak conditions. For example, numerous prospective studies have shown a link between fresh and marine water exposure from swimming and gastrointestinal illness. Although there has been no documented scientific evidence that the current Part 503 has failed to protect public health, the National Research Council (NRC) committee on biosolids stated that "additional scientific work is needed to reduce persistent uncertainty about the potential for adverse human health effects from exposure to biosolids." They recommended the use of "improved risk-assessment methods to establish better standards for chemicals and pathogens." (NRC, 2002).

In many types of environmental exposures, such as those associated with biosolids, there is often limited ability to measure health risks directly. This is due, in part, to the limited sensitivity of epidemiology studies. The field of microbial risk assessment was developed to help characterize relatively low-level risks due to exposure to environmental contamination by pathogens. The information provided by risk assessments has been used for both operational and regulatory purposes. For example, risk assessment has been used to evaluate the need for wastewater treatment improvements (Soller et al., 2003) and has also provided information on appropriate levels of treatment to maintain tolerable risk levels (Regli et al., 1991b, U.S. EPA, 2006). Risk assessment also can be a useful decision-making tool, especially when comparing one decision with another (Eisenberg et al., 2002). For example, two management practices can be assessed side by side to evaluate which is the lower risk option and whether the difference is significant. Alternatively, risk assessment can be used when a decision must be made on whether or not a utility should modify a treatment process.

It is important to have clear expectations of what risk assessment can and cannot do. On one end of the spectrum, there is a view that risk assessment can provide precise estimates of risk that result in a definitive answer to a regulatory or management problem. On the other end of the spectrum there exists the view that environmental processes are too complex for a risk model to be of any use. In reality, neither view is correct. Environmental and infection processes are sufficiently complex so that precise estimates are not currently possible given our current state of knowledge on these processes. Analytical and modeling techniques, however, are available that allow for risk estimates to provide sufficient and valuable information for decision making and for identifying important data gaps that can direct future research needs. For example:

- A risk model can be used to assess whether one exposure is a greater concern than another;
- A risk model can assess which factors (parameters) contribute the most to the uncertainty in the risk estimates; i.e., collecting better data on which factor will give us the most improvement on our risk estimates;
- A risk model can be used to identify the type of monitoring/surveillance strategy that will provide the best information to estimate risk; and
- A risk model can assess how changes in a treatment process or in environmental factors will affect the risk.

Each of these can provide regulators and/or managers with valuable information. This information, however, must be used in a context that acknowledges the uncertainty and variability inherent in the environmental process.

In this report a risk assessment approach is presented to estimate the health effects associated with the exposure of pathogens from biosolids application. The following sections of Chapter 1.0 provide an overview of the approach (further described in Chapters 2.0, 3.0, and 4.0). First the exposure component is framed. Each component of the exposure pathway is described along with the specific data needs. Next, the health effects component is framed. Here the researchers discuss the difference between a static individual-level risk approach, that borrows from a chemical risk paradigm, as well as a dynamic population-level risk approach that accounts for the unique aspects of infectious disease processes.

1.2 Intended Audience

The research summarized in this project reports will be helpful for a wide range users including:

- Regulators (local, state, and federal);
- Biosolids producers (engineers, operators);
- Environmental managers (Treatment facility managers, city council members, etc.); and
- Concerned citizens

For regulators the microbial risk assessment methodology demonstrated in this project can provide a tool to set risk-based standards for microbial contaminants in biosolids, in a manner similar to the way that chemical risk assessment iis used for regulation and management of commercial and industrial chemicals and hazardours wastes.

Biosolids producers, engineers and operators would benefit if this risk-based approach were used as a common metric to develop a general consensus for evaluation of microbial health risks associated with biosolids management. Similar to current experience with accepted chemical risk assessment methodology, a general consensus regarding the microbial risk framework can help biosolids producers interpret monitoring results and evaluate the potential health-risk benefits of making operational changes to existing treatment and/or adding additional treatment processes. Similarly, environmental managers will be better able to make informed decisions about potential capital improvements, land application site placement issues, and/or other related questions.

Finally, this research provides a mechanism and metric for technical professionals, regulators, and the general public to discuss and compare the relative risks associated with land application of biosolids, both as a general practice and in specific, actual or proposed instances.

1.3 Defining a Risk-based Approach for Exposures to Pathogens

The two fundamental components of a microbial risk assessment are exposure and the health effects (International Life Sciences Institute et al., 1996). The exposure assessment estimates the quantity of pathogens (dose) to which individuals or populations are exposed. Exposure modeling can take advantage of environmental models that describe the mechanistic process of fate and transport, as well as empirical relationships based on pathogen or surrogate indicator data. Often a combination of monitoring data and modeling is used to estimate the dose to which a population is exposed. The dose is then input into a health effects model that estimates the risk associated with a specific pathogen/exposure scenario. The risk characterization step then examines risk estimates for different pathogen/exposure scenarios to provide an overall risk summary. Part of this summary allows for the estimation of incremental increased risk from biosolids, above background risk.

This risk assessment process for exposure to pathogens via a biosolids route of exposure is shown schematically in Figure 1-1. The initial five boxes represent the environmental processes that are involved in producing an exposure dose. An overview of these processes is provided in the next few sections.

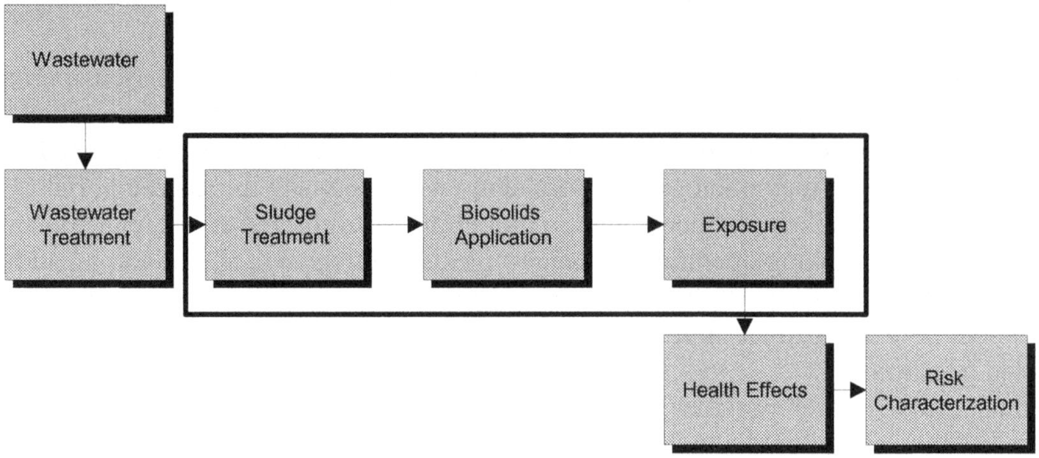

Figure 1-1. Schematic Diagram of Risk Assessment Process.

1.3.1 Framing the Exposure Component

As noted above, exposure assessment is generally concerned with estimating the dose of pathogens to which individuals may be exposed. That dose is then used as the input variable for a health effects model. In the area of microbial risk assessment, two exposure routes that have been extensively studied are drinking water and recreational water. For drinking water exposures the exposure assessment has generally focused on characterizing site-specific source water concentrations of pathogens (often *Giardia*, *Cryptosporidium*, and enteric viruses), and fitting these data to a statistical distribution. Some level of drinking water treatment efficiency is assumed and an exposure concentration is estimated. From this exposure concentration, a water consumption rate is used to estimate the dose. Water consumption rates have generally been based on data from either Rosebury and Burmaster (1992) or more recently from a U.S. EPA report (U.S. EPA, 2000). For recreational water exposures, concentrations of indicator organisms at the recreational water site are monitored. Depending on where the monitoring sites are in comparison to the swimming sites, fate and transport hydrology models have been used to estimate the concentration of indicators at the exposure sites. These data are then used as input into concentration-response curves. These concentration-response curves are generated epidemiology studies that collect both indicator data at the swimming site and illness data from the swimmers.

Exposure assessment for biosolids is similar in some aspects to recreational water and drinking water exposures and different in other aspects. Like drinking water exposures, pathogen levels in biosolids may be below detection limits. To estimate pathogen levels at the site of exposure requires measuring pretreated sources and assuming a level of removal due to the treatment processes. An exception to this for drinking water might be the few public supply wells that are not disinfected. In these cases it is sometimes possible to measure viruses in untreated ground water and calculate risk directly. Like recreational water exposure assessments, fate and transport models are sometimes needed since the site of exposure may not be where biosolids are applied. There are a number of exposure routes including aerosolization, runoff to surface waters, and movement into groundwater that would require a prediction of the fate and transport from application site to exposure site. Unlike recreational water and drinking water, there is one primary source of contamination, sludge from wastewater treatment plants, and unlike drinking water, biosolids exposures are intermittent depending on how often they are applied to a specific site, and their proximity to an exposed population..

After the wastewater is treated, one of its by products is sludge. Sludge becomes the initial product introduced to the next three processes shown in Figure 1-1: sludge treatment; biosolids application; and exposure. These processes will be central to the exposure scenario definitions discussed in Chapter 2.0, where the following is defined: the sludge treatment processes that are currently being used, the methods and location that the biosolids are being applied, and the important exposure pathways that humans may potentially be exposed to pathogens that originate from biosolids. The primary data needs for a risk assessment are environmental measurements of pathogens. Often environmental measurements at the site of exposure do not detect pathogens. This is true for treated drinking water. Drinking water regulations, therefore, are based on source water pathogen data coupled with specific treatment requirements.

Likewise, biosolids monitoring data seldom identify pathogens. In fact, by definition Class A biosolids can never have detectable levels of viruses. In part, due to this lack of detectable pathogen data, the regulations developed by the U.S. EPA are operationally based. This operational standard required that the pathogens in biosolids be reduced to below detectable levels for Class A and significantly reduced, but detectable levels, for Class B. The specific regulation was thought to be protective of public health.

An operationally based regulation, however, is not based directly on risk. Lack of illness data may be due to the fact that there are no appreciable risks, or it may be due to the fact that sufficiently sensitive studies have not been conducted and/or these studies are difficult to implement. Likewise, monitoring data below detectable limits do not necessarily imply zero risk. Because of this limitation in interpreting the data, and because measurements of biosolids below detectable limits do not provide sufficient information on risk, this report presents a more scientifically rigorous microbial risk approach to estimating risks associated with biosolids exposure. Though this approach is applied to biosolids the method also has broader application to environmental exposure of pathogens.

1.3.2 Framing the Health-effects Component: I. Individual-level Assessment

Using post-treatment pathogen data to estimate risk suggests the following risk model.

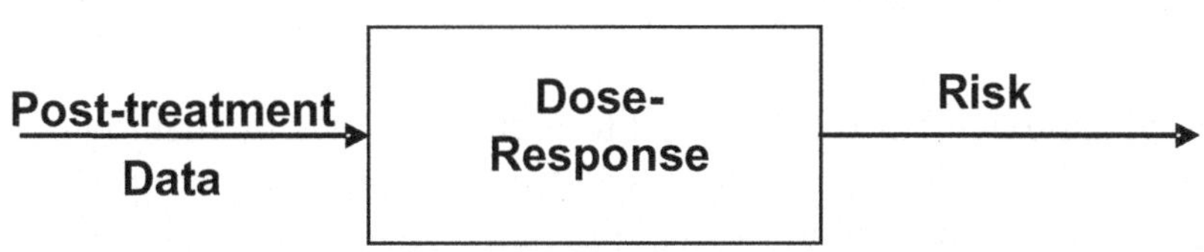

Figure 1-2. Risk Model Using Post-Treatment Pathogen Data.

If measurable pathogen data exit in biosolids, a traditional risk assessment approach would be to use these data as input into a dose-response function (a function that converts a pathogen concentration into a probability of infection or illness) to estimate risk. However, available post-treatment monitoring data consist of mostly values below detectable levels.

Observations with values below detectable levels mean that with a specified level of confidence, one can be certain that the concentration of pathogens is below a specific limit. In addition, since only a small sample of biosolids is actually tested, a test result that is below the detectable level does not mean that the concentration of pathogens is below the detection limit value for the entire biosolids pile or for biosolids piles that have accumulated over time. Often monitoring consists of infrequent testing of small samples that are presumed to be representative of much larger masses of biosolids.

One possible approach to estimating risk is to make assumptions about the meaning of a nondetectable value. At one extreme one can assume that measurements below detectable levels do represent the fact that there are no pathogens in the sample tested and that the sample is representative of the entire biosolids pile; i.e., the biosolids pile is completely homogeneous. Under these assumptions the risk associated with exposure to biosolids in which monitoring data consist of values below detectable levels is zero. Another extreme that can be assumed is that all values below detectable levels represent pathogen concentrations at the detection limit; i.e., if the detection limit is 1 PFU/4 grams then all samples contain 1 PFU/ 4 grams of virus. The researchers can also assume that this sample is representative of the entire biosolids sample, i.e., the biosolids pile is completely homogeneous, or one can assume that there may be parts of the biosolids pile that have detectable concentrations of pathogens. This illustrates the potential for a biased result depending on the assumptions used in interpreting the data.

The above discussion suggests that monitoring data consisting of all or mostly values below detectable levels (at < 1 PFU/4 gm) do not provide sufficient information to obtain unbiased estimates. An alternative approach is to move upstream in the biosolids process and take advantage of the raw sludge data. This is especially vital for Class A since by definition all measurements are nondetectable; it is also valuable for Class B since it can inform the decision on the impact of operational changes. The new risk model can be depicted as follows.

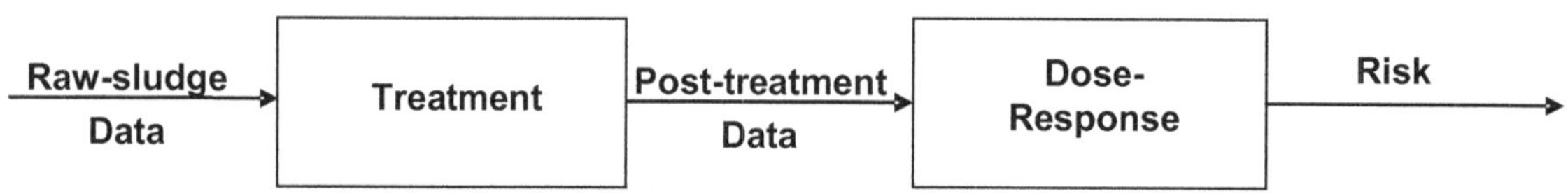

Figure 1-3. Risk Model Using Raw Sludge Pathogen Data.

This model-based approach has the potential to reduce the bias by making defensible assumptions about the biosolids treatment processes and subsequent exposure pathways, and by allowing the data to inform and constrain our risk estimates. Virus concentrations from raw sludge serve as an additional data source to the post-treatment data, as does the data on treatment process efficacy. When post-treatment data contain more nondetectable values, the raw sludge is relied on more heavily in the risk calculations. As post-treatment data contain more detectable data, the raw sludge data are relied on less in the risk calculation. In this manner, data from different stages of the biosolids process can be used to minimize the biases in risk estimates. This approach to using upstream data to help estimate risk can be framed as a Bayesian problem in which the raw sludge and treatment data are considered prior information that can inform the risk estimate, and the post treatment data are used to inform the likelihood. As with any Bayesian problem, if the data informing the likelihood are sparse, the posterior (or risk estimate) is strongly influenced by the prior information. As the data improve, the posterior (or risk estimate) is influenced less by the prior information. Chapter 4.0 presents this approach in more detail illustrating its use through a hypothetical example.

The next section examines two approaches for estimating health risks using an individual-based approach, which focuses on risks due to environmental exposure to an individual, and a population-based approach, which focuses on the effects of environmental exposure on disease incidence in a population.

1.3.3 Framing the Health-effects Component: II. Population-level Assessment

Chemical risk assessment has been used extensively since the NRC report on risk assessment in 1983 (NRC, 1983). By comparison, microbial risk assessment methodologies are still under development. Because of the earlier development of chemical risk assessment, attempts to provide a quantitative assessment of human health risks associated with the ingestion of pathogens have generally focused on calculating the probability of individual infection or disease as a result of a single exposure event (static models) (Haas, 1983, Regli et al., 1991a). The U.S. EPA realized that infectious disease processes had features significantly different from disease processes caused by chemical exposures and began preliminary work by sponsoring a workshop in 1996, from which a working group developed a microbial risk framework (International Life Sciences Institute et al., 1996). In contrast to the previous framework, the U.S. EPA microbial risk framework encourages the consideration of pathogen specific epidemiological data, such as incubation period, immune status, duration of disease, and rate of symptomatic development. However, the ILSI/U.S. EPA framework falls short of explicitly requiring the consideration of features specific to the disease transmission process that would be necessary to highlight the distinction between infectious disease risk and a more conventional static risk process (Soller et al., 1999).

There are two fundamental features of infectious disease transmission processes that are different from chemical agent risk. First, pathogens are transmitted from host to host either directly or indirectly via the environment. The principal feature of this process is that all pathways of transmission are integrated, one influencing another in such a way that they cannot be considered independent processes. For example, a child may be infected due to pathogen exposure from environmental pollution and transmit the infection through person-to-person contact in household, school, or day care settings. In this example, the person-to-person transmission is dependent on and the result of environmental exposure. Second, after initial exposure to a pathogen, a host may exhibit a certain degree of protection from disease (immunity). The degree to which a pathogen is circulating within a population and the extent to which the population has previously been exposed to the pathogen determines the degree to which protection is conferred on the population.

Phase I of this investigation involved the development of a microbial risk assessment methodology/framework that explicitly encompasses these features (integration of transmission pathways and presence of immunity), in which biosolids were included as one of the potential routes of exposure to pathogens (Colford et al., 2003; Eisenberg et al., 2004). In this section, a brief overview is provided of the health effects model that was developed during phase I of this investigation.

1.3.3.1 Static Risk Assessment Model

Microbial risk assessment involves evaluating the likelihood that an adverse health effect may result from human exposure to one or more pathogens. Assessments using a static model (NRC, 1983) for evaluating microbial risk typically focus on estimating the probability of infection or disease to an individual as a result of a single exposure event. These assessments generally assume that multiple or recurring exposures constitute independent events with identical distributions of contamination. Secondary transmission and immunity are assumed to be negligible or that they effectively cancel each other out. Secondary transmission would increase the level of infection/disease in a community relative to a specific exposure to pathogens, and immunity would decrease the level of infection/disease in a community relative to a specific exposure to pathogens (Eisenberg et al., 2004).

In the static model, it is assumed that the population may be categorized into two epidemiological states: a susceptible state and an infected or diseased state. Susceptible individuals are exposed to the pathogen of interest and move into the infected/diseased state with a probability that is governed by the dose of pathogen to which they are exposed and the infectivity of the pathogen. A schematic diagram of a static model is presented in Figure 1-4.

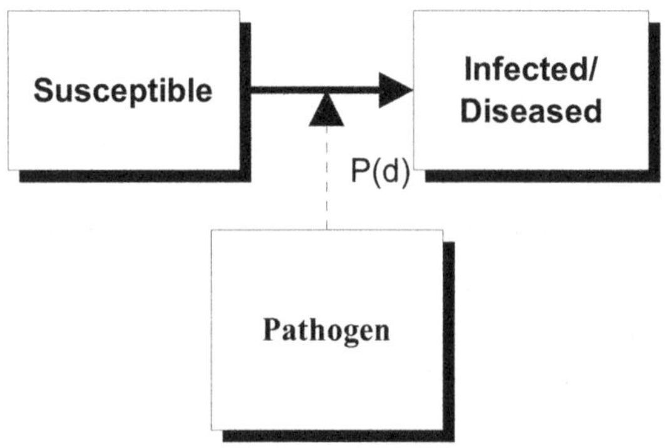

Figure 1-4. Static Risk Assessment Conceptual Model.

The probability that a susceptible individual becomes infected or diseased is a function of the dose of pathogens to which that individual is exposed. When individuals in a susceptible state are exposed to pathogens from an environmental source, they move with a given probability to an infected or diseased state. This probability dose-response function is labeled P(d) in Figure 1-4. The dose is typically calculated by estimating two quantities: the concentration of pathogens at the exposure site and the volume (of water, or in this case biosolids, or biosolids amended soil) ingested. This dose quantity is then input into a dose-response function that estimates the probability that an exposed individual will become infected or diseased.

The critical health effects information required for the static model, therefore, is summarized in the pathogen-specific dose-response function. The probability of infection following exposure to a virulent pathogen depends on several host- and pathogen-specific factors. The interaction between a pathogen and the host can be viewed as a series of conditional events, in which each event must occur in order to result in infection. The infection status depends on a number of factors such as: 1) the number of organisms that enter the host; 2) the host's ability to inactivate these organisms; 3) the number of organisms that can withstand the host's local immune defenses, adhere to mucosal surfaces, and multiply in order to infect the host; and 4) variation in pathogen virulence and host susceptibility (Colford et al., 2003, Eisenberg et al., 1996, Eisenberg et al., 2004).

Infection may be defined as the invasion, colonization, and multiplication of a pathogenic microorganism (Teunis et al., 1996). Since humans may come into contact with a wide range of pathogen concentrations, it is therefore useful to develop a functional relationship between a quantitative measure of microorganism ingestion and the risk of infection. The probability of infection is often multiplied by the number of exposed individuals to estimate the expected number of infected individuals for the exposure scenario under consideration.

1.3.3.2 Dynamic Risk Assessment Model

In a dynamic risk assessment model, the population is assumed to be broken into a more comprehensive series of epidemiologically relevant states. Individuals move from state to state at rates that are based on clinical data such as the duration of infection, duration of immunity, etc. Only a portion of the population is in a susceptible state at any point in time, and only those in the susceptible state can become infected or diseased through exposure to the pathogen of interest. The probability that a susceptible person moves into an exposed state is governed by the dose of pathogen to which they are exposed, the infectivity of that pathogen, and also the number of infected/diseased individuals with whom they may come into contact.

For both dynamic and static representations of the disease process, infectivity as a function of dose (estimated using a dose-response function) is an important factor in estimating risk. The dose-response function is the critical component of a static microbial risk assessment model. The dose-response function is also important in a dynamic microbial risk assessment model; however, other factors such as person-to-person transmission, immunity, asymptomatic infection, and/or incubation period may also be important.

Accounting for these additional factors when estimating risks associated with exposure to pathogenic microorganisms requires a more sophisticated mathematical model than the static model shown in Figure 1-4. When a dynamic disease transmission model is used, one can account for attributes specific to the transmission of infectious diseases. Depending on the infectious disease processes that are important for a specific exposure, the dynamic model may include more or fewer components, and therefore vary in complexity.

To model exposure to pathogens via a biosolids route of exposure, a dynamic model was developed during phase I of this investigation. The model accounts for person-to-person transmission, immunity, incubation, and asymptomatic infection (Figure 1-5). Similar to the static model, the solid lines in Figure 1-5 represent the movement of individuals from one epidemiological state to another, and dotted lines represent the movement of pathogens. For a more complete discussion of this health effects model, please refer to the Phase I final report (Colford et al., 2003).

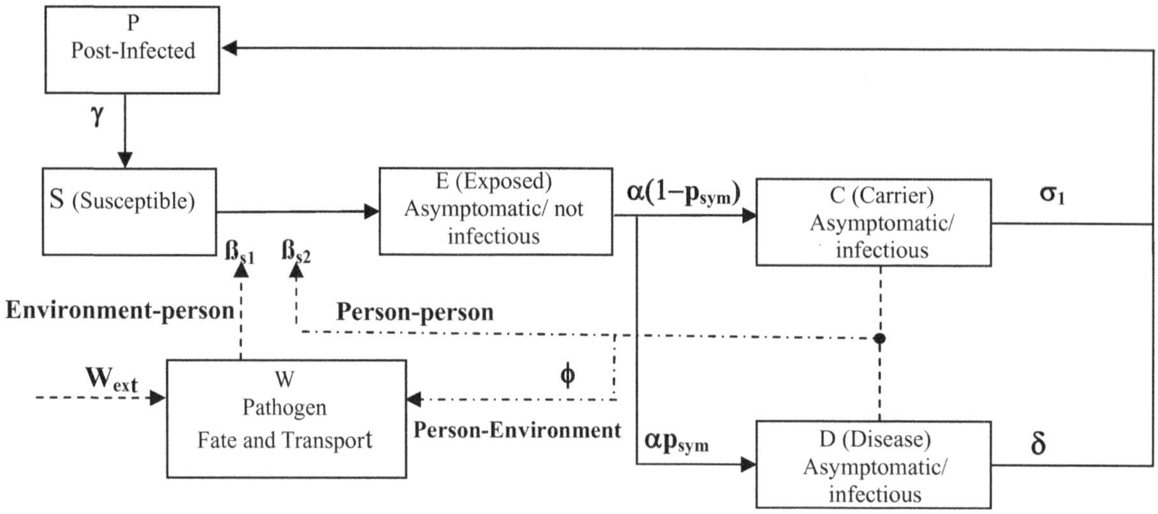

Figure 1-5. Conceptual Model for Health Effects. Solid lines represent the movement of individuals to different disease states, whereas the dotted lines represent the movement of pathogens. S, E, C, D, and P represent numbers of individuals, and W represents number of pathogens in the environment.

1.4 Roadmap of Report

This chapter presents a motivation for using a risk-based approached and provides a brief overview of both an individual and population-level framework for risk assessment that integrates three primary data sources (pathogen concentrations in raw sludge; treatment process efficacies; and post-treatment pathogen concentrations in biosolids) in order to obtain risk estimates.

The remainder of this report further develops the risk-based approach presented in these introductory sections. Chapters 2.0 and 3.0 provide additional details of the framework. Chapter 2.0 defines the exposure scenarios and the available data needed to specify the model. Chapter 3.0 presents the environmental pathway models used in this report: groundwater and aerosolized pathogens. Chapter 4.0 defines specific models and presents the risk estimates for these models using different treatment options and environmental pathways. Chapter 5.0 provides a discussion of the simulation results and on the future direction of microbial risk assessment applied to biosolids related exposures.

For those less interested in the technical aspects of the analysis, some of the chapters and text in this report can be skipped. For example, Chapter 2.0, Chapter 3.0, and Figure 1-1 of Chapter 4.0 provide technical details on our risk model and the data. The report is written so that the reader can choose to skip these sections and focus on Chapters 1.0, 4.0, and 5.0.

CHAPTER 2.0

EXPOSURE SCENARIOS

Data necessary to characterize the exposure can come from a number of different environmental processes along the pathway from wastewater to biosolids application, and during the fate and transport of organisms to an exposure site (See Figure 1-1). At each point along this environmental pathway two types of data can be collected. The first is monitoring data that measures the occurrence of pathogens. Monitoring data can be collected using different temporal and spatial sampling protocols. The second type of data are fate and transport data. Fate data describe the die-off of pathogens as they move through different natural or treatment processes, whereas transport data describe the transit of organisms from one point in the process to another. All of these data are highly dependent on environmental factors such as temperature, moisture content, pH, etc.

When conducting a risk assessment it may be possible to use different combinations of these data. For example, monitoring data may describe the concentration of a particular pathogen at a specific biosolids application site at (or prior to) the time that an individual is exposed. Such data would be useful for characterizing the exposure dose. Depending on the sampling strategy used, however, these data may not be sufficient to capture the variability in pathogen concentrations due to a variety of climatic and environmental factors. Designing an appropriate sampling strategy is important in capturing this variability. For example, high pathogen concentrations may occur 1% of the time. A sampling strategy that only samples once per month may never catch these high concentration events resulting in biased risk estimates. For cost or logistic reasons, data collected at a sufficient resolution may not be attainable. An alternative solution is to take advantage of data sources upstream from biosolids, such as pathogens in raw sludge, and then to model the exposure process.

From a risk assessment perspective, models are useful because they can help capture the variability due to known environmental factors such as temperature, and they allow for the exploration of alternative conditions in the treatment process or in environmental conditions or provide a means to generalize results to different locations. In this manner risk models can be used to predict a change in risk given a change in a particular set of environmental conditions. For example, a manager may want to estimate the health benefits associated with increasing the holding time of biosolids prior to application, or they may want to know if a larger setback distance from residential housing will decrease the risk of exposure through aerosol transport. A regulator, on the other hand, may want to estimate the health benefits associated with different regulations. For example, what should the setback distance be for biosolids application? Will one setback regulation be appropriate for all applications? Or they may want to be able to estimate how the variability in environmental processes will result in periodic extreme risk events. All of these issues can be addressed through risk models, given that appropriate data are available to parameterize the model.

Models do have their shortcomings. Given the limited data and large uncertainties inherent in environmental processes, models will introduce variability into the risk estimates. Variability can be introduced either through inaccuracies in model specification or in parameter identification. The more complicated the model, the more variability that is introduced. To minimize variability it is important to develop functional relationships between environmental processes and environmental factors that are supported by data. For example, to effectively predict the difference between composting in different times of the year there needs to be a functional relationship between temperature and pathogen die-off rate within the compost.

These model requirements dictate specific data needs. Monitoring and surveillance may not provide the appropriate data for reasons discussed above. Rather, carefully designed experimental studies that examine how processes vary by environmental conditions are important in providing data that can be used to develop such functional relationships. Another way to minimize variability is to estimate a risk difference between one condition and another. This comparative risk estimate will have much less variability then the two absolute risk estimates used to calculate the risk difference.

In general, even when the models are carefully developed, there exists a tradeoff between using models to correct for potential biases that may be due to reliance on sparsely collected monitoring data and the variability introduced by the models themselves. The following sections in this chapter describe the different components of an exposure scenario and the specific data needs.

2.1 Fate and Transport Information Needed to Define an Exposure Scenario

The details of the exposure model for a risk assessment depend on how the exposure assessment is approached. As previously discussed, relying on monitoring data at a certain point in the process may preclude the need for understanding the fate and transport prior to that process point. This concept is illustrated in Table 2-1. The first column contains the different process points along the environmental pathway. Each row then summarizes the fate and transport information required, given that monitoring data are available for that process point. For example, if there existed reliable pathogen data at the location of human exposure, the exposure model would simply be the frequency and duration of exposure (amount ingested, duration and frequency of swimming, etc.). If there were reliable monitoring data post treatment, an exposure model would include environmental conditions prior to and after the biosolids are applied, as well as the environmental variables associated with the particular exposure pathway of interest (direct ingestion, ground water infiltration, surface water runoff, aerosol transport, etc.).

A more detailed description of the types of data relevant to characterize an exposure model is summarized in Table 2-2. As in Table 2-1, the first column in Table 2-2 lists the process

Table 2-1. Data Matrix for Characterizing Exposure.

	Data Specifications Required for an Exposure Assessment						
Environmental process point where monitoring data are available	Infection incidence data	Type of treatment	Treatment conditions	Environmental conditions prior to biosolids application	Environmental conditions post biosolids application	Environmental variables for exposure pathway	Human exposure information
Raw wastewater	X	X	X	X	X	X	X
Treated wastewater		X	X	X	X	X	X
Treated biosolids				X	X	X	X
Time of application					X	X	X
Time of exposure						X	X
Location of exposure							X

points shown in Figure 1-1. The second column describes the monitoring site that would provide pathogen concentration data, the third column describes the point where there may be pathogen reduction data available, and the fourth column provides a more detailed list of data needs and the different environmental process points. For biosolids treatment, the list of dewatering and stabilization methods is meant to include the important treatment processes and is not meant to be exhaustive. In any given scenario a subset of these methods may be used. Temperature, time of process, final percent moisture, and the presence of chemical additives are listed as important environmental factors that help characterize the pathogen reduction associated with a given treatment process. The data associated with biosolids applications will be method and location dependent, but in general will be a function of time, temperature and moisture. The exposure pathways each have their environmental factors associated with fate and transport of the pathogen to a given exposure type. As will be discussed later, the focus for this project will be on three pathways: groundwater, aerosol, and direct ingestion.

For the purposes of this investigation, defining an exposure scenario is comprised of defining and characterizing the following:

- The pathogen
- The population group
- The exposure pathway
- The characteristics of the biosolids application
- The specific biosolids treatment characteristics.

The types of attributes that each of those components may take on for a specific exposure scenario are described briefly below.

2.1.1 Pathogen

Pathogens present in biosolids are primarily the same pathogens found in municipal wastewater. The primary source of pathogens in the raw wastewater is from human sewage. The profile of pathogens, therefore, reflects the profile of infections within the community and the concentrations are dependent on the shedding rate of infectious individuals. Nonpoint sources may also enter into the wastewater stream resulting in some pathogens from other animal hosts. Therefore, the mix of pathogens potentially in biosolids is comprised those organisms that are important waterborne pathogens. The viruses include adenoviruses, astroviruses, noroviruses, rotaviruses, Hepatitis A, Hepatitis E, and enteroviruses. The bacteria include *Campylobacter*, pathogenic *E coli* pathotypes such as 0157:H7, *Salmonella*, and *Shigella*. The protozoa and helminths include *Cryptosporidium*, *Giardia*, and *Ascaris*.

While not routinely obtained, there are limited data on a number of pathogens in raw sludge and biosolids, such as enteroviruses, *Salmonella*, *E. coli*, *Giardia*, *Ascaris* and *Cryptosporidium*. Most pathogen monitoring, however, is associated with Class A samples and are therefore predominately nondetectable values. Routine monitoring is limited to fecal coliform and *Salmonella,* and to a much lesser extent, enteroviruses, and viable helminth ova. No such monitoring is routine in raw sludge.

Table 2-2. Expanded Data Matrix for Characterizing Exposure.

	Data relevant for exposure assessment		
Process Point	**Monitoring Site**	**Pathogen Reduction Site**	**Data Needs**
Wastewater	Pathogen shedding levels or Incidence of infection in population		
Wastewater Treatment	Raw wastewater		
Biosolids Treatment	Treated wastewater/sludge	Aerobic digestion Anaerobic digestion Composting Lime stabilization Lagoon storage Heat drying	Log-removal of processes
		Factors affecting treatment reduction	Temperature Time Final % moisture Chemical addition
Biosolids Application	Biosolids	Storage (prior to application)	Storage time Storage temperature % Moisture at application Moisture during storage Pathogen survival rate
	Time of application	Application site	Time to exposure Temperature prior to exposure Addition of moisture prior to exposure Pathogen survival rate via exposure route

		Exposure site	
Exposure	Drinking water	Via groundwater exposure	Soil type Depth to groundwater Travel time/distance Aquifer characteristics Water ingestion rate Application rate
	Aerosols	Via aerosol exposure	Wind speed Distance to population Duration of exposure Transport characteristics of pathogen (size shape quantity) Temperature and thermal gradients Sunlight uVA and uVB intensity Bioaerosol conc. in nearby areas Time of day / time of year Relative humidity Application rate
	Application site	Via direct ingestion exposure	Amount of soil ingested Application rate
Health Effects	Exposure dose or disease incidence attributable to the exposure		

2.1.2 Population Group

Two primary types of exposure to pathogens via biosolids are likely to occur: occupational and residential. Occupational exposure occurs when individuals are exposed to pathogens as a result of their vocation. Residential exposure could occur when individuals are exposed to pathogens from land-applied biosolids due to the proximity of their residences to a biosolids application site. For example, residential exposure could occur under conditions where pathogens were transported via wind from a biosolids application site to a community located nearby.

With respect to the population group that is potentially exposed, another consideration is whether the assessment is to be conducted for adults or children. For example, if it is assumed that individuals are exposed to biosolids amended soil through the direct ingestion of biosolids amended soil, it is possible that children and adults would be exposed to different quantities of that soil.

2.1.3 Potential Exposure Pathways

Results from previous biosolids risk assessment work carried out by U.S. EPA suggest that the most important biosolids related exposures can be classified into five pathways (U.S. EPA, 1991a, U.S. EPA, 1991b, U.S. EPA, 1992a). Those five pathways are summarized as follows:

- Direct ingestion of biosolids-amended soil;
- Ingestion of surface water containing microorganisms from erosion of land-applied biosolids;

- Ingestion of groundwater containing microorganisms via leaching through soil from land applied biosolids;
- Ingestion of plants grown in biosolids amended soil; and
- Exposure to (ingestion or inhalation of) wind transported microorganisms from land applied biosolids dispersed into the air.

The characterization of these exposure pathways requires monitoring data, fate and transport data and predictive models. Direct ingestion, which was addressed in Phase I, is the simplest pathway to characterize. Other pathways such as groundwater are much more complex.

2.1.4 Biosolids Application

Biosolids can be applied to "non-public contact sites" such as agriculture lands, forests, or reclamation sites or "public contact sites" such as parks, plant nurseries, golf courses, lawns, etc. (U.S. EPA, 1993a). There are public access restrictions in place for each scenario and are set at 30 days and 1 year respectively. For risk assessment, one important consideration is that the rate at which biosolids are applied differs by type of application. For example, reclamation rates of biosolids application are substantially higher than agronomic (agricultural) rates. The geographic location and characteristics of the application site are also potentially important factors that could impact human exposure to microorganisms in biosolids. Geographic location accounts for factors that may change from site to site such as soil type, and weather (temperature, wind conditions, etc.).

Biosolids are generally applied by spraying, spreading onto the surface, disking, or injecting below the surface. Biosolids can be applied in either a liquid or dewatered state. The specific method used will be one determining factor in deciding which exposure pathways are important.

2.1.5 Biosolids Treatment

The treatment process is focused on reducing the contaminant concentration and pathogen levels in the sludge. Sludge consists of a primary component, the portion of wastewater solids contained in the raw influent that is removed in the sedimentation process, and an activated sludge component that comes from secondary treatment processes. The type of biosolids treatment that is employed at a particular location as well as the characteristics of that treatment will affect the magnitude of the pathogen concentration in any subsequent exposure to the biosolids. The general types of pathogen treatment that could be employed to treat biosolids are typically categorized into the class of biosolids that they produce (Class A and Class B). Class A biosolids may have to meet specific process controls to insure that pathogens are below detectable limits. Possible processes include composting, heat drying, heat treatment, thermophilic aerobic digestion, irradiation, and pasteurization. Class B biosolids, for which certain management practice requirements are imposed, may have pathogens above detectable limits but must be at reduced levels. Potential processes include aerobic digestion, air-drying, anaerobic digestion, composting, and lime stabilization.

The treatment of sludge to obtain a biosolids product involves a number of steps. Each step can involve a number of different processes depending on a variety of environmental and economic considerations. A possible treatment regime may employ the following three steps. The first step combines the sludge components from different points along the wastewater treatment process, specifically the primary and waste activated sludge components. This step may also involve thickening or blending which is meant to increase concentrations, but has no effect on pathogen levels. The second step is generally concerned with reducing vector attraction and pathogen concentrations. This step can consist of: 1) anaerobic or aerobic digestion, which provide biochemical decomposition of organic matter into carbon dioxide and either methane gas if anaerobic, or water if aerobic. The decomposition is accomplished by microorganisms and will reduce pathogen concentrations; 2) lime treatment, which elevates pH, releases ammonia gas, controls odor and reduces pathogens; 3) heat treatment, which kills pathogens; and 4) composting, an aerobic process, which provides organic solids, pathogen, and odor reduction. Typically, this step is limited to one of these four processes but there are occasions where, for example, digestion is followed by lime or heat treatment. If necessary, the third step may involve sludge conditioning followed by dewatering. Dewatering can be accomplished using a variety of equipment including, vacuum filters, centrifuges, drying beds, filter presses, belt filters. The purpose of this last step is to reduce volumes and make it easier to transport the biosolids. The amount of dewatering, typically from 15 - 30% solids, will not affect pathogen levels.

The environmental conditions for each of these treatment processes, such as temperature, time, moisture content, and pH, ultimately determines its efficacy. Some of the treatment processes, such as anaerobic digestion or heat treatment, are highly controlled and not affected by external environmental conditions. Other treatment processes, such as aerobic digestion, composting, and lime treatment, are less controlled and may be affected by external conditions and must be properly design, operated and maintained to minimize the variability in their treatment efficacy. It is therefore important to incorporate into the exposure assessment model, to the extent possible, pathogen reduction rates as a function of these design considerations.

2.2 Data Availability

A literature review and data collection effort were conducted. The purpose of these efforts was to determine what data were available in the scientific literature and from unpublished sources (from municipal agencies) that could be of use for identifying and characterizing representative exposure scenarios for use in this phase of the investigation. The following is a summary of these efforts.

2.2.1 Literature Review

The goal of the first step in the literature review was to identify the relevant published articles. This was carried out using the University of California (electronic) library system. Articles collected during the first phase of this investigation were also reviewed. A total of 206 articles were collected. Of these articles, approximately 173 were identified as potentially relevant for the exposure component of this investigation. The main focus of those articles is summarized in Table 2-3.

Table 2-3. Literature Review Overview.

Literature Review Summary	#
Treatment	
Monitoring and Surveillance (m/s)	23
Fate and Transport	6
Application	
Monitoring and Surveillance (m/s)	5
Fate and Transport	32
Exposure	
Aerosol (F/T)	7
Groundwater (F/T)	56
Soil Ingestion	16
Lab	6
Miscellaneous	22
Total	173

A large portion of the articles (101) focused on the fate and transport of pathogens through the environment. Of the remaining 72 articles, 28 discussed monitoring or surveillance results (concentrations of pathogens at various stages of the biosolids generation process), 16 discussed soil ingestion, six discussed laboratory methods for detecting pathogens or the relative efficiencies of various laboratory methods. For the purposes of this investigation, 22 articles were classified as miscellaneous. The majority of these were either review articles or only tangentially related to the exposure scenario portion of this investigation.

The 144 articles focusing on monitoring/surveillance, fate/transport and soil ingestion were of most interest in identifying and characterizing potential exposure (excluding the laboratory and miscellaneous articles). The relationship between these articles and the important environmental processes/components of the exposure model development discussed previously (Table 1-1) are presented in Figure 2-1.

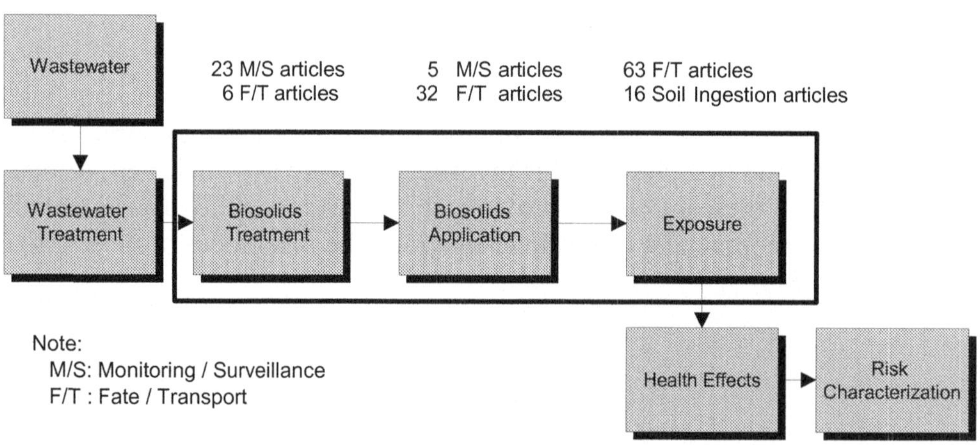

Figure 2-1. Important Environmental Processes in Exposure Model Development.

Another useful way to categorize the articles is by pathogen. Table 2-4 separates the articles into two categories: the first contains all articles with virus and possibly bacteria data and the second category contains articles with only bacteria data. The following sections summarize the 68 articles that contain virus data (See Appendices A and B for a complete list of these 68 articles; 47 are listed Appendix A and 21 focusing on inactivation rates in groundwater in Appendix B).

Table 2-4. Literature Review by Pathogen.

	Articles Containing Virus Data	Articles with only Bacteria Data	Total
Treatment			
Monitoring and Surveillance	4	19	23
Fate and Transport	6	0	6
Application			
Monitoring and Surveillance	0	5	5
Fate and Transport	15	17	32
Exposure			
Aerosol (F/T)	6	1	7
Groundwater (F/T)	37	18	55
Soil Ingestion	-	-	16
Total	68	60	144

2.2.1.1 Treatment

There are two main steps in the biosolids treatment process: dewatering, such as air drying, and stabilization, such as aerobic and anaerobic digestion, composting and lime stabilization. All articles in this treatment section address one of these two treatment steps.

2.2.1.1.1 Monitoring/Surveillance

The first of the four monitoring/surveillance treatment articles relating to viruses (Gerba et al., 2002) provides concentrations of enteroviruses in mesophilic anaerobically digested biosolids at a treatment facility in Arizona. A total of 13 samples were taken over the period of approximately three years and measured for enterovirus concentrations in MPN/4 grams (Table 2-5). The detection limit of the technology used to measure the viruses is as low as 0.2 MPN/4 grams.

This article also contains fate and transport data, providing an estimated log-reduction of enteroviruses during air drying. The article also provides similar estimated log-reductions during aerobic and anaerobic digestion, composting and lime stabilization. The article does not provide any other environmental information about the specific processes from which these estimates were obtained.

Table 2-5. Enterovirus Concentrations in Mesophilic Anaerobically Digested Biosolids.

Sample Date	Treated Biosolids Enteric Virus (MPN/4 grams)
Dec-96	0.2
Feb-97	1.3
Jul-97	1.3
Oct-97	1.38
Dec-97	1.18
Mar-98	0.24
Jul-98	60.09
Nov-98	2.77
Feb-99	0.17
Jun-99	0.5
Aug-99	0.54
Dec-99	2.38
Mar-00	0.55

The second article (Hamparian et al., 1985) provides the number of samples containing isolated viruses in both aerobic and anaerobically digested biosolids. Limited information concerning the treatment plants sampled is provided.

The third article (Soares et al., 1994) provides the concentrations of enteroviruses in sludge before and after mesophilic anaerobic digestion.

The fourth article (U.S. EPA, 1988) provides occurrence of viruses and bacteria in treated sludge as well as inactivation rates of viruses in soil for different soil conditions and temperatures and for various types of treated sludge. This article also contains fate and transport data, providing an estimated log-reduction for mesophilic anaerobic digestion, aerobic digestion, composting air drying and lime stabilization.

2.2.1.1.2 Fate and Transport

Four of the six fate and transport articles related to viruses focus on treatment while the remaining two focus on storage. The first treatment article (Sorber and Moore, 1986a) provides log-reductions of viruses during air-drying, aerobic and anaerobic digestion, composting and lime stabilization. The second article (Goddard, 1981) provides the percent reduction of viruses during aerobic and anaerobic digestion. The authors reported process information, such as temperature, length of process in days and moisture content. The third article (Ward and Ashley, 1977) provides data on the effect of dewatering on the inactivation rates of enteric viruses in seeded raw sludge. The fourth article (Hurst, 1989) provides an equation for inactivation of viruses per day based on incubation temperature.

The first (Sanin et al., 1994) of the two storage articles lists the log-reductions of both aerobically and anaerobically digested sludge samples after they were frozen at three different rates, three different temperatures and three different times and subsequently thawed. The second article (Ahmed and Sorensen, 1995) gives the log reduction for poliovirus for temperatures ranging from 5-50°C.

2.2.1.2 Application

No monitoring/surveillance data of viruses at application sites were found in the literature. All published application data collected (15 articles) were concerned with fate and transport.

Three of the 15 articles focused on adsorption of viruses to different soil types. The first article (Sobsey et al., 1980) provides data on the percentage of poliovirus and reovirus adsorbed for 8 different soil types, the effect of pH on virus adsorption, and the log-reduction in the various soil types. The second article (Goyal and Gerba, 1979) discusses virus adsorption to sandy loam. The third article (Burge and Enkiri, 1978) compares five soil types by examining the type of adsorption isotherm and rates of adsorption.

Two articles provided data on the sorption of bacteriophages and viruses on clays and soils (Chattopadhyay and Puls, 1999a, Chattopadhyay and Puls, 1999b).

Five articles provided data on the survival of viruses in different soil types. The first article gives the percent removal of viruses in various types of soil (garden soil, sandy loam, dune sand) (Gerba et al., 1975). The second article provides first order die-off rates of viruses and their half-lives (Reddy et al., 1981). The third article gives log-reductions of poliovirus and indicator viruses in sand, muck and clay at a specific temperature (Meschke and Sobsey, 2003). The fourth article provides log-reduction of viruses in soil, sandy and loam, sterile and non-sterile (Bagdasaryan, 1964). The fifth article provides viral reduction and inactivity in soil saturated with secondary effluent and the effect of soil microbial activity on virus survival (Nasser et al., 2002). One article is related to agriculture and provides data on viral persistence both in soil and on lettuce and radishes grown in vegetable pots under natural conditions. The soil was flooded with poliovirus inoculated raw waste, activated sludge, or secondary effluent that was obtained from a municipal activated sludge plant (Tierney et al., 1977). Environmental data were provided such as temperature and rainfall.

One article provides concentrations of viruses in soil cores of fine sand treated with lagooned sludge (2/3 anaerobic and 1/3 aerobic) at specific time points (Bitton et al., 1984). Soil temperature, moisture and rainfall were also provided.

The remaining three articles provide data on: 1) how natural organic matter influences transport of viruses in sand and soil (Zhuang and Jin, 2003a); 2) factors that affect pathogens through soil (Mawdsley et al., 1995); and 3) the effect of different surfactants and organic matter on bacteriophages (Chattopadhyay and Puls, 2002).

2.2.1.3 Exposure

Articles were categorized in this group by three types of exposures: aerosol, groundwater and soil ingestion.

2.2.1.3.1 <u>Aerosol</u>

Two of the six articles focusing on aerosol exposure provide models for estimating downwind concentrations of viruses from an aerosol source (Brooks et al., 2005, Dowd et al., 2000). Two articles provide general information on bioaerosols and public health (Pillai, 2002, Pillai and Ricke, 2002). One article focuses on estimation of the emission rate of pathogens from land applied biosolids (Teltsch and Katzenelson, 1978). The last is a review article concerning the fate and generation of bioaerosols (Brooks et al., 2004).

2.2.1.3.2 Groundwater

Of the 37 articles related to groundwater exposure, 21 give inactivation rates in groundwater at specific temperatures with some limited information about the conditions of the water (e.g. filtered/sterile). A WHO monograph (Pedley et al., 2006) had summarized the data from 19 of these 21 articles (see Appendix B for more details). The two other estimates were obtained from a personal communication from Dr. P. Berger (U.S. EPA, HAC member). These 21 articles are listed in our database but the articles were not collected. Table 2-6 provides the summary of the 87 different estimates of inactivation rates of viruses provided by these 21 articles. The inactivation rates are shown for various temperatures and water conditions. For example, the inactivation rate of Coxsackievirus at 10°C in sterile water is 0.019 (day^{-1}) (Matthess et al., 1988).

Table 2-6. Literature Review by Pathogen.

Virus	# Inactivation Rates
Coxsackievirus	10
Echovirus	11
Hepatitis A	6
Poliovirus	26
Rotavirus	1
Simian Rotavirus	2
Bacteriophage	31

Eight of the remaining 16 articles related to groundwater provide models for fate and transport of viruses in porous media (Bhattacharjee et al., 2002; Faulkner et al.; 2002, Faulkner et al., 2003; Schijven and Hassanizadeh, 2000a; Schijven et al., 2002; Schijven and Hassanizadeh, 2000b; Sim and Chrysikopoulos, 2000). The remaining nine articles provide data for the following topics: survival of viral pathogens and indicators in chlorinated water; survival and transport of bacterial pathogens and indicator viruses in sandy aquifers; filtration and deposition profiles in saturated quartz sand; groundwater transport of viruses; comparison of saturated and unsaturated flow under leaching conditions similar to land treatment; attachment and detachment rates for viruses; comparison of pathogenic viruses and MS2 in terms of removal by sand and effects of Aluminum-oxide on transport of viruses (Allwood, 1993; Deborde et al., 1999; Dowd and Pillai, 1997; Lance and Gerba, 1984; Redman et al., 2001; Schijven et al., 2003; Schijven et al., 2000; Schijven and Simunek; 2002, Zhuang and Jin, 2003b).

2.2.1.3.3 Soil Ingestion

There are 16 articles that document the levels of soil ingestion in children and adults (Binder et al., 1986b; Calabrese, 1991; Calabrese and Stanek, 1995; Calabrese et al., 1997a; Calabrese et al., 1997c; Clausing et al., 1987b; Finley et al., 1994; Hawley, 1985b; LaGoy, 1987b; Simon, 1998; Stanek and Calabrese, 2000; Stanek and Calabrese, 1995; Stanek et al., 1997; Stanek et al., 2001; Thompson and Burmaster, 1991; U.S. EPA, 1989a).

Of these 16 articles, five provided data relevant for estimating daily soil ingestion estimates for the purpose of identifying and characterizing exposure scenarios (Table 2-7). Most recent soil ingestion estimates are based on tracer studies. In general for all of these studies, individuals are followed over a period of time and daily stool samples collected from each participant. One or more tracer elements such as aluminum, silicon and titanium, are then measured in feces, based on the premise that these elements are poorly absorbed by the gastrointestinal tract. Based on the amounts of tracer elements isolated from the feces and on the concentration of these tracer elements found in the soil, ingestion estimates are then calculated for each tracer element.

The assumptions in this method are that: 1) tracer intake by other routes (food) is low and not too variable; 2) tracer adsorption in the gastro-intestinal tract is low; and 3) tracer concentrations in soil are high and not too variable.

One issue that researchers had to contend with was a high inter-tracer measurement variability. This high variability presents significant interpretational problems when concerned with choosing which tracer provides the most accurate soil ingestion estimate. A methodology denoted Best Tracer Method (BTM) was developed to reduce misalignment and improve estimated precision of tracer recovery (Calabrese et al., 1997b).

Two of the earlier studies were conducted in the mid 1980s. Hawley et al. (1985a) produced estimates without using tracers by measuring the dirt on hands and observing how often hand were put in ones mouth. Binder et al. (1986a) estimated high levels of ingestion that not even the authors considered accurate as the motivation was to present a new method of estimating soil ingestion and not to accurately provide soil ingestion estimates. This method involved estimating the amount of soil ingested by ruminants to measure the amount of soil ingested by children. The following year, Clausing et al. (1987a) used feces tracer levels from six hospitalized children, assumed to have no soil ingestion, as a background level. They then measured tracer levels in 18 children visiting a nursery school and subtracted out the background level. The resulting ingestion rate was 56 mg/day. Ten years later Calabrese et al. (1997b) conducted a similar study in 64 children residing on a superfund site in Montana. Their estimate, 26.6 mg/day, was half of that found in the Clausing et al. (1987a) study. Finally, Lagoy et al. (1987a) provided age-stratified estimates of soil ingestion using both the Binder and Clausing data. Table 2-7 provides a summary of these studies on soil ingestion.

Table 2-7. Soil Ingestion Literature Summary.

Reference	Type of Study	Soil Ingestion Estimate	Notes
Binder et al., 1986	Tracer - aluminum, silicon and titanium	-Aluminum and silicon estimates: 181 and 184 mg/day respectively. -Titanium estimate: 1,834 mg/day.	-59 children aged 1-3 years old from Montana - The authors do not consider these estimates accurate, but they do consider the method reasonable. -Lagoy 1987 suggests that children may have been ingesting titanium from another source.
Calabrese and Stanek, 1997a; 1997b	Tracer + BTM – 8 elements	Mean ingestion for children was 26.5625 mg/day.	-64 children 1-4 years old residing on superfund site in Montana. - 7 day follow up period. -Food and fecal samples analyzed on daily basis. -Conducted in the Fall (issues of a season effect)
Clausing et al., 1987	Titanium, aluminum and acid insoluble residue -Combine results of 3 tracers for an average ingestion	Average estimated soil ingestion rate: amongst the nursery school children was 105 mg/day, and amongst hospitalized children was 45 mg/day. -Using hospitalized children as background, average daily soil ingestion is 56 mg/day.	-18 children visiting a nursery school and 6 hospitalized children without any possibility of soil contact. -Children aged 2-4 years old. -Study period was 5 days.
Hawley, 1985	-Children's ingestion estimates based on amount of dirt found on hands and how often hands put in mouth -Adult estimate based on size of their hands	- 2.5 yr olds: 250 mg/day. - 6 year olds: 50 mg/day -Adults (8 hours of yard work 2 day per week): 480 mg/day.	-Used estimates from Lepow et al. who hypothesized a 250mg/day ingestion rate with as much as 10mg ingested from a child putting his/her fingers in his/her mouth a single time. Measured soil by removing dirt from hands with an adhesive -A second study Roel's et al. estimated 40-180mg of soil on one hand of 11 year old. -Adults assumed to ingest twice daily a quantity of soil corresponding to one half the covering of the inside surface of the fingers and thumbs of both hands
Lagoy, 1987	Uses 2 pilot studies (Binder et al. and van Wijnen and Clausing)	-1-6 year olds: 100 mg/day -6-11 year olds: 50 mg/day -Adults (over 11) with frequent hand to mouth contact (e.g. smokers) or who are in direct contact with soil (construction workers, gardeners): 50 mg/day -Adults under normal conditions: 25 mg/day	Maximums: -1-6 year olds: 500 mg/day - 6-11 year olds: 250 -Adults (over 11): 100 mg/day

2.2.2 Monitoring Data From Utilities

A survey was developed and distributed to determine if relevant data for this phase of the investigation were available in unpublished sources. A total of 52 responses was obtained from agencies in the U.S., Canada, and Australia. Seven respondents reported that their agency collects or has collected enterovirus data.

Of the seven respondents, enterovirus data from raw sludge and post treatment were obtained from four utilities. At each of the utilities, samples were taken over a period of 1-3 years and analyzed for enteric virus levels, reported in Plaque Forming Units (PFU) per 4 grams total solids. Three of the four utilities also measured other pathogen levels such as fecal coliform, salmonella, and helminth ova. A summary of the data obtained from the utilities are provided in Table 2-8.

Table 2-8. Summary of Data Provided by Utilities.

Location	Time Period	Sampling Frequency	Pathogens Monitored	Treatment	Monitoring Data for Treated Biosolids?	Annual Biosolids Production
Dataset 1	07/2001 - 07/2004	Monthly	Fecal Coliform Enteric Virus Helminth Ova	Anaerobic Digestion, belt press	Yes	48,000 dry tons
Dataset 2: Site 1	10/1994 - 4/1997	22 samples taken in time period - not at regular intervals	Salmonella Enteric Virus Helminths	Anaerobic Digestion (35°C±2°C) Lagoon Aging, Stablization and Dewatering Air-Drying	Yes Data for all 3 stages of treatment	190,000 dry tons
Dataset 2: Site 2	10/1994 - 3/1997	24 samples taken in time period - not at regular intervals	Salmonella Enteric Virus Helminths	Anaerobic Digestion (35°C±2°C) Lagoon Aging, Stablization and Dewatering Air-Drying	Yes Data for all 3 stages of treatment	
Dataset 3	6/2001 - 6/2003	7 samples taken in time period - not at regular intervals	Salmonella Fecal Coliform Enteric Virus Helminth Ova	3 Stage Theromophilic Anaerobic Digestion	Yes After 1st and 2nd stages of treatment	10,000 dry tons
Datsset 4	1/2003 - 12/2004	24 samples taken monthly	Enteric Virus	2 Stage Thermophilic Anaerobic Digestion	Yes After treatment and prior to farm application	68,000 dry tons

2.2.2.1 Dataset 1

The first of the utilities, denoted here Dataset 1, provided three full years of monthly data (Table 2-9). Samples for raw blended sludge were taken from a sample port on the line pumping blended sewage sludge. The sample consists of approximately 40% primary treatment sludge, 50% waste activated sludge and 10% anaerobically digested primary sludge. For analysis, two 1000 ml plastic containers were normally cooled and shipped UPS next day air to the processing laboratory. Two 1000 ml containers of raw sludge were used to provide enough sludge at approximately 4% total solids to conduct the enteric virus analyses. The standardized method used to analyze the concentration of viruses within a sample has a detection limit of 1 PFU/4 grams. The detectable monthly concentrations (greater than 1 PFU/4 grams) are estimated to the nearest integer.

Table 2-9. Dataset 1: Raw Sludge Monitoring Data From Utility 1.

Sample Date	Raw Sludge Enteric Virus (PFU/4 grams)	Treated Biosolids Enteric Virus (PFU/4 grams)	Sample Date	Raw Sludge Enteric Virus (PFU/4 grams)	Treated Biosolids Enteric Virus (PFU/4 grams)	Sample Date	Raw Sludge Enteric Virus (PFU/4 grams)	Treated Biosolids Enteric Virus (PFU/4 grams)
Aug-03	<1	<1	Aug-02	<1	<1	Jul-01	<1	<1
Sep-03	<1	<1	Sep-02	<1	<1	Aug-01	<1	<1
Oct-03	<1	<1	Oct-02	7	<1	Sep-01	<1	<1
Nov-03	<1	<1	Nov-02	<1	<1	Oct-01	<1	<1
Dec-03	<1	<1	Dec-02	<1	<1	Nov-01	<1	<1
Jan-04	5	<1	Jan-03	<1	<1	Dec-01	<1	<1
Feb-04	2	<1	Feb-03	<1	<1	Jan-02	<1	<1
Mar-04	1	<1	Mar-03	<1	<1	Feb-02	<1	<1
Apr-04	1	<1	Apr-03	<1	<1	Mar-02	<1	<1
May-04	7	<1	May-03	2	<1	Apr-02	<1	<1
Jun-04	6	1	Jun-03	222	<1	May-02	<1	<1
Jul-04	31	<1	Jul-03	<1	<1	Jun-02	<1	<1

2.2.2.2 Dataset 2

The second of the utilities, denoted here Dataset 2, provided data for two sites over a period of 2.5 years (Table 2-10). Over the time period 22 samples of from site 1 and 24 samples from site 2 were analyzed. Samples of 200 ml of raw sludge were taken from the digester feed. Samples of size 200 ml were also taken from the digester feed on the same days as the digester draw.

At each site, two different treatment processes are used simultaneously. In the first process, anaerobically digested material is conditioned with a cationic polymer, and dewatered by centrifugation to a cake with solids content 20-30%. The cake is aged in lagoons to stabilize the sludge, it is then air-dried in batches to ~60% solids content. In the second process, anaerobically digested material is pumped to lagoons for dewatering, aging and stabilization. It is then air-dried in batches to ~60% solids content. Samples were taken at the lagoon and air-drying steps for each treatment process at each site. Samples were taken from the lagoon at 1,2,4,6,9,15,18 months and samples were taken throughout the air drying process (until it reached 60% solids). Sample sizes from the lagoon and air-drying processes were as large as 1000 g wet weight.

Enteric viruses were enumerated according to the method described in the Annual Book of ASTM standards (ASTM, 2005). The detection limit for viruses was 1 PFU / 4 grams for this method. However, the detection limit was sometimes lowered to 0.0051 PFU / 4 grams by increasing the sample mass used for analyses.

Table 2-10. Dataset 2: Raw Sludge Monitoring Data From Utility 2.

Dataset 2: Site 1

Sample Date	PFU/ 4 gram total solids
10/31/1990	12.0000
11/30/1990	6.0000
4/18/1991	1.0870
5/2/1991	4.7337
6/6/1991	0.5814
7/18/1991	19.4444
8/15/1991	0.1000
9/12/1991	<1
10/10/1991	<1
11/28/1991	1.0290
1/9/1992	0.1889
2/13/1992	0.5422
3/5/1992	0.2689
4/9/1992	9.2664
4/22/1992	2.7027
6/11/1992	0.7505
7/16/1992	2.5641
8/20/1992	1.1905
9/24/1992	10.0000
10/22/1992	7.8947
11/19/1992	2.5424
1/7/1993	1.3514

Dataset 2: Site 2

Sample Date	PFU/ 4 gram total solids
10/31/1990	15.0000
11/30/1990	2.0000
4/18/1991	1.7758
5/2/1991	1.0270
6/6/1991	1.8727
7/25/1991	10.0000
8/15/1991	1.1000
9/12/1991	1.0000
10/10/1991	1.0000
11/28/1991	0.8259
2/13/1992	0.4562
3/5/1992	0.0905
4/9/1992	4.9383
4/22/1992	1.3267
6/11/1992	0.9685
7/16/1992	1.4184
8/20/1992	14.6341
9/24/1992	4.8780
10/22/1992	10.2941
11/19/1992	7.8431
1/7/1993	1.2232
3/4/1993	1.3333
3/4/1993	2.2727

2.2.2.3 Dataset 3

A total of seven samples of raw sludge were taken from a gravity thickener over a period of two years (Table 2-11). The treatment has both a two- and three-stage thermophilic digestion process. The first stage is mesophilic digestion with one digester with an average detention time of 2.5 days at 95°F and is the same for both the two- and three -stage processes. For the two-stage process, the second stage is thermophilic digestion and consists of two digesters where the first has an average detention time of 11 day days at 126°F and the second has an average detention time of 9.4 days at 120°F. The second stage of the 3-stage process is one of two digesters that both have average detention times of 16 days at 125°F. The third stage of the three-stage process consists of one digester with an average detention time of five days at 115°F.

The biosolids are dewatered with a belt filter press to 19% solids, and hauled to composting operations. They are not certified as Class A as they leave the plant.

Estimated enterovirus concentrations from samples are provided at both the end of the two-stage and three-stage processes.

Table 2-11. Dataset 3: Raw Sludge Data From Utility 3.

Sample Date	PFU/ 4 gram total solids
6/14/2001	102
6/20/2001	36
6/27/2001	46
9/10/2001	58
10/9/2001	18
11/5/2001	10
12/3/2001	34

2.2.2.4 Dataset 4

Raw sludge samples were taken monthly over a year and measured for enteric virus levels (Table 2-12). The treatment is a two-stage thermophilic digestion process. The first stage digesters operate at 128°F for 12.0 days (continuous flow). The batch stage operates at 128°F for 16 hours. The biosolids are centrifuged to 30% solids. They are then fed to silos and from the silos into trucks. An average of 700 wet tons/day is hauled to a farm that is 120 miles away. Enterovirus concentrations estimated from samples are provided from raw sludge, from the silos and prior to farm application.

Table 2-12. Dataset 4: Raw Sludge Data From Utility 4.

Sample Date	PFU/ 4 gram total solids
1/7/2003	10
2/4/2003	10
3/4/2003	15
4/2/2003	9
5/8/2003	13
6/3/2003	34
7/1/2003	51
8/5/2003	48
9/2/2003	36
10/7/2003	19
11/4/2003	40
12/2/2003	18
1/7/2004	19
2/10/2004	2
3/9/2004	2
4/6/2004	7
5/4/2004	21
6/8/2004	59
7/13/2004	102
8/10/2004	62
9/7/2004	36
10/5/2004	16
11/2/2004	30

2.3 Exposure Scenario Characteristics

The literature reviews described above were carried out to identify data that could be used to characterize representative and realistic exposure scenarios in which individuals are assumed to be exposed to microbial pathogens associated with land applied biosolids.

This section describes characteristics of the exposure scenarios for this investigation. This identification process was based on the data identified and presented in the previous section, conversations and meetings with the WERF project subcommittee (PSC) and health advisory committee (HAC), the relative intensity of different types of biosolids treatment practices in the United States, and the potential public health importance related to different exposure scenarios and pathways. Brief descriptions of and the rationale for the exposure scenario characteristics that were investigated during this phase of the investigation are provided below.

2.3.1 Population Group

The population group of primary interest for this risk assessment methodology demonstration is residential. The PSC and HAC members recommended unanimously to focus on a residential population, using occupational exposure for comparison. The issue behind this decision was that although occupational groups were in general more highly exposed, they were also in general not representative of different susceptible population subgroups. Residential groups both represented a larger exposed population size and included susceptible subgroups such as children, immunocompromised, the elderly, etc. The risk assessment methodology during phase I of this investigation, however, could be employed to characterize the risk associated with occupational exposures. In Chapter 4.0, assessments are performed for occupational workers for the ingestion pathway.

2.3.2 Pathogen Choice

Based on the review of available data, the two most likely candidates for the risk assessment methodology demonstration are *Salmonella* and enteroviruses. Both pathogens are currently monitoring options for Class A biosolids determination. They both satisfy the following criteria for selecting a model pathogen: 1) Occurrence data must be available for the pathogen in biosolids; 2) Human dose-response data must be available for the pathogen; and 3) The pathogen must be of potential concern in biosolids.

Based on discussions and meetings with the PSC and HAC, it was agreed that this phase of the investigation should focus on just one pathogen, allowing for a more thorough exploration of the exposure scenarios. It was subsequently agreed that enteroviruses would be more appropriate than *Salmonella* based on the following: 1) enteroviruses are more readily transported via aerosols and groundwater; and 2) enteroviruses do not experience regrowth (as *Salmonella* does) post-biosolids treatment. Regrowth is a complicated and not well documented phenomenon and therefore was thought to be beyond the scope of this project. Again, it should be noted that the risk assessment methodology could be employed to carry out an assessment of the risk associated with *Salmonella* via biosolids exposure.

2.3.3 Biosolids Treatment

The scenarios for this project used pathogen concentration data from untreated sludge and model the biosolids treatment process (in terms of pathogen reduction). Specifically, mesophilic anaerobic digestion and lime treatment were modeled.

2.3.4 Biosolids Application

Based on conversations and meetings with PSC and HAC members, the methods of biosolids land application that are of primary interest, based on the relative level of use intensity in the United States include surface application, disking, and injection into the ground. For the purposes of the risk assessment methodology demonstration, the focus of this phase of the investigation was on surface sapplication. PSC and HAC members agreed that this is a realistic and reasonable approach, particularly given its relevance to the aerosol pathway.

2.3.5 Exposure Pathway

For the purposes of this risk assessment methodology demonstration, the focus of this phase of the investigation was on 1) the direct ingestion of biosolids amended soil, 2) ingestion of groundwater containing microorganisms via leaching through soil from land applied biosolids, and 3) exposure to (ingestion or inhalation of) wind transported microorganisms from land applied biosolids dispersed into the air.

An exposure model for the direct ingestion of biosolids amended soil was used in the case study example during phase I of this investigation. That same model was employed for demonstration purposes during Phase II.

For exposure to (ingestion or inhalation of) wind transported microorganisms from land applied biosolids, the risk assessment methodology demonstration adapted recently published models (Brooks et al., 2005, Dowd et al., 2000).

For ingestion of groundwater containing microorganisms via leaching through soil from land-applied biosolids, the project team collaborated with HAC member Dr. P. Berger to identify an appropriate groundwater transport model. PSC and HAC members agreed that pathogen transport through karst soils was of highest interest. Therefore, a simple groundwater model appropriate for demonstrating the risk assessment methodology with this type of exposure was developed.

CHAPTER 3.0

EXPOSURE PATHWAY MODELS

This chapter outlines the modeling strategy for the groundwater and aerosol exposure pathways. The goal of the modeling process is to define models with sufficient detail to capture the complexity of the fate and transport processes, without being overly computationally intensive. In addition to being computationally intensive, complex models can be more difficult to interpret. In a risk assessment often a simpler environmental model is more advantageous than a complex one given the relatively low level of precision of risk estimates.

The output of the groundwater models is a distribution of attenuations. These attenuations are used in Chapter 4.0 to estimate risks associated with the groundwater exposure pathway. The output of the aerosol model is a distribution of downwind concentrations. Both the groundwater and aerosol exposure pathway models serve to add additional attenuation of pathogen concentrations in the overall exposure models. The following groundwater and aerosol exposure models are described by the most significant processes and parameters according to published literature and provide the level of detail necessary for the estimation of relative risk in the overall risk assessment.

3.1 Groundwater Exposure Pathway

Groundwater passes through three distinct types of media: non-porous media (karst/bedrock), porous media and both saturated and unsaturated soil. The viruses from biosolids applied to land travel through one or more of these types of media. To simplify the exposure modeling process, three scenarios were identified for virus transport in groundwater:

1) Karst/Bedrock (Non-porous media). Fractures in the karst result in free flow of pathogens. Since no data on attenuation in this type of media were found, a 1 - 3 log removal of viruses was assumed (Phil Berger, HAC member, personal communication). This log removal is the sum of two processes:
    i) Travel time to well (~1 d) results in roughly 0.5 log removal.
    ii) Adsorption results in 1-1.5 log removal.
2) Unsaturated soil followed by non-porous media
3) Unsaturated soil followed by saturated soil

Instead of using an extremely complex model that allows for heterogeneous media, simpler models for each media type were used. These simpler models assume homogenous media. Each media type is modeled separately, and the resulting attenuation in each media type is summed to obtain the overall attenuation for the heterogeneous scenario. For example, scenario 3 uses one model to estimate the attenuation in unsaturated media and a second model to estimate the attenuation in saturated media. The two estimates are then summed to obtain a total attenuation for this heterogeneous scenario.

The following three sections describe the models used for the three media: karst, unsaturated and saturated.

3.1.1 Non-porous Media (karst/bedrock)

The assumption of the 1 – 3 log-removal in non-porous media does not consider a setback distance, which is consistent with the fact that there is no federal regulation specifying the need for a setback. Information collected from various states however, suggests that many require some form of setback depth to groundwater and/or bedrock (see Appendix C for requirements by state). Therefore, a 1 – 3 log removal may underestimate the actual log removal for states with a setback requirement.

3.1.2 Unsaturated Zone

Attenuation in unsaturated soil depends on such factors as thickness of unsaturated layer (*L*), hydraulic conductivity (K_s), sorption (κ°) and inactivation (λ, λ^*). See Table 3-1 for specific definitions of these factors.

The Virulo software package (Version 1.0, http://www.epa.gov/ahaazvuc/csmos/models/virulo.html) is used for predicting virus attenuation in unsaturated soils. Virulo outputs a probability of failure of achieving a pre-specified level of attenuation. For the purposes of this risk assessment, a model for the attenuation of viruses, not the probability of attenuation of viruses is required. Thus, the Virulo output could not directly be used. Instead the underlying model and parameter values specified by the Virulo software were used to generate distributions of attenuation.

To estimate die-off, Virulo models the partitioning of viruses to the liquid-solid and air-water interface. An important factor in estimating die-off is virus sorption within unsaturated porous media and is significantly affected by the presence of these two interfaces. Figure 3-1 gives an illustration of viruses in the various interfaces as they are distributed through an unsaturated medium.

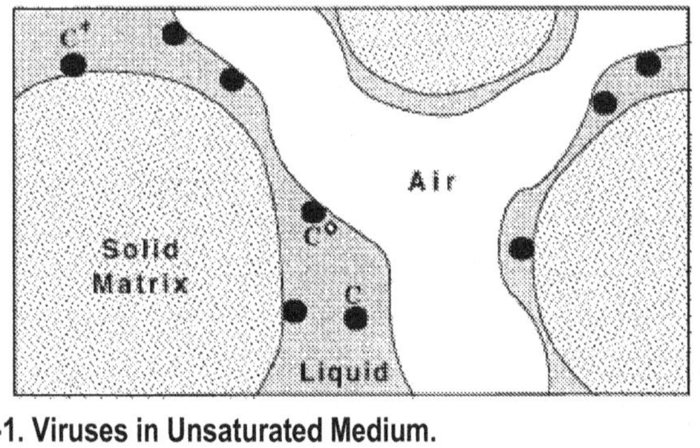

Figure 3-1. Viruses in Unsaturated Medium.

C represents viruses in the liquid phase, C^* represents viruses sorbed at liquid-solid interface, and C° represents viruses sorbed at air-liquid interface.

The governing equations for the fate and transport of viruses in unsaturated soils, as well as a closed-form solution to these equations, were obtained from the Virulo documentation (see Appendix D for details). The solution provides an equation for an attenuation factor *A*, where $-\log_{10} A$ is the log attenuation of active viruses.

Virulo was also used to obtain parameter values since it contains a database of default parameters for various soil types. Table 3-1 lists the parameter values required to calculate the attenuation factor and the source of these values. These parameters can be classified into two groups: hydraulic (flow) and virus parameters.

Table 3-1. Parameter Values Required to Calculate the Attenuation Factor.

Parameter	Definition	Reference	Notes
Flow Parameters: Values vary by soil type			
θ_r	residual soil water content	Virulo	
θ_m	moisture content	Virulo	
θ_s	water content of a saturated porous medium	Virulo	
K_s	saturated hydraulic conductivity	Virulo	
α	water retention curve fitting parameter	Virulo	
n	curve fitting parameter	Virulo	Well tabulated, empirical
ρ	bulk density of the soil	Virulo	
r_p	average radius of soil particles	Virulo	
α_z	vertical dispersivity	Virulo	
Virus Parameters: Values vary by virus			
λ	inactivation rate coefficient of viruses in the liquid phase	Virulo	
λ^*	inactivation rate coefficient of viruses sorbed at the liquid-solid interface	Virulo	
κ	liquid to liquid-solid interface mass transfer coefficient	Virulo	
κ°	liquid to air-liquid interface mass transfer coefficient	Virulo	
r_v	radius of the virus	Virulo	
K_d	equilibrium distribution coefficient	Virulo	
Constants			
k_b	Boltzmann's constant	Constant	1.380658×10^{-23} J K^{-1}
μ	viscosity of water	Varies with temperature	Tables with values by temperature
ρ_w	density of water	Varies slightly with temperature	~1 g/mL
g	acceleration due to gravity	Constant	9.8 m/s^2
σ	surface tension of water	Varies with temperature	Tables with values by temperature

3.1.2.1 Virulo's Hydraulic Parameter Defaults

Virulo provides default values (mean and standard deviation) for 11 soil types: clay, clay/loam, loam, loamy/sand, sand, sandy/clay/loam, sandy/loam, silt, silt/loam, silt/clay, silt/clay/loam. By selecting a particular soil type, the flow parameters in Table 3-1 are automatically populated with default values. The hydraulic property defaults, e.g. K_s, K_d, κ are estimated unsaturated hydraulic properties from surrogate soil data such as soil texture, e.g. sandy or loamy, and bulk density, ρ. In this way, basic soil data are translated into hydraulic properties (called pedotransfer functions). This program begins with class average hydraulic parameters for each USDA soil textural class and builds on these estimates using neural network models. The mass transfer parameters were obtained from (Chu et al., 2001).

3.1.2.2 Virus Parameter Defaults

Virulo provides virus parameter defaults (inactivation rates, λ and λ^*, mass transfer coefficients, κ and $\kappa°$, virus radius, r_v, and the equilibrium distribution coefficient, K_d) for the following viruses by soil type:

- Polio-clay
- Polio-silt
- Polio-sand
- Hepatitis A – clay
- Hepatitis A – sand
- Reovirus 3 – clay
- Reovirus 3 – silt
- Reovirus 3 – sand
- Coxsackievirus – clay
- Coxsackievirus – sand
- Echovirus – clay
- Echovirus – silt
- Echovirus – sand

The basis for selecting these values is given in Faulkner (2002).

3.1.3 Saturated Zone

The steady-state solution to a one-site kinetic model was used to estimate the removal of viruses in saturated soil. The model is based on attachment and detachment rates, inactivation rates of free and attached viruses, dispersivity, velocity, and distance. Under steady-state conditions, the relative contributions of inactivation and adsorption to the removal of viruses by soil passage can be computed analytically. A steady-state situation occurs when input of virus is continuous and may be seen as a worst-case situation (Schijven and Hassanizadeh, 2000b). The steady-state solution to the kinetic model is provided in Appendix E.

The model parameter estimates were obtained from a column experiment (Schijven et al., 2002). The column experiments were designed to be similar to field conditions, and were thus carried out a low temperatures (5 ± 3°C) at which inactivation rates are low. Two columns were prepared; the first column was operated at a pore water velocity of 1.6 meters per day and the second at 3.5 meters per day. The model parameters were estimated for both columns.

Both one-site and two-site kinetic model parameters were estimated by fitting the breakthrough curves. To fit the one-site model, the values for site 2 were set to zero. At low temperatures, it was found that interaction with site 1 accounted for almost all the removal of viruses. Thus, when estimating the overall removal of viruses, the one-site model is an adequate model. That is, the additional complexity of the two-site model does not improve the overall estimate of removal.

The dispersivity and velocity were calculated directly from the experimental conditions. The remaining four parameters were estimated from fitting breakthrough curves. Point estimates as well as 95% confidence intervals were provided for the attachment and detachment rate coefficients and the inactivation rate coefficient of attached viruses, whereas a point estimate was provided for the inactivation rate of free viruses. The parameter estimates from the second column with a pore water velocity of 3.5 meters per day were selected as they estimated virus log removals similar to previous studies. In one field study, concentrations of MS2 and PRD1 were reduced by 8 $\log_{10}$ over 30 meters (Schijven and Hassanizadeh, 2000b). Another study estimated concentrations of MS2 and PRD1 were reduced by 6 and 7.5 $\log_{10}$ over 40 meters respectively (Deborde et al., 1999). Using the mean parameter estimates from column 2, the one-site model predicts a 7.5 log removal over 30 meters.

3.1.3.1 Virus Inactivation

An important parameter in the saturated model is virus inactivation. Temperature is the most important factor in virus inactivation. Several simple linear equations in temperature have been published.

The literature collection provided 88 estimates of inactivation rates by temperature of viruses in groundwater from various publications (Appendix B). Using simple linear regression of the inactivation rates on temperature, the following equation for the prediction of inactivation was obtained:

$$\hat{\lambda} = 0.042T - 0.121$$

where $\hat{\lambda}$ is the predicted attenuation and T is the temperature in degree Celcius.

The following 2 equations of inactivation rate equations were found in the literature:

$$\lambda = 0.018T - 0.144 \text{ (Yates et al., 1985)}$$

and

$$\lambda = 0.033T - 2.4 \text{(Schijven and Hassanizadeh, 2000b).}$$

These equations were only used for comparison purposes and were not used as the parameter values in the model. Instead, the default Virulo parameter estimates were used.

3.1.3.2 Model Virus

Model viruses are needed to represent the behavior of pathogenic viruses during subsurface transport and to predict their removal. MS2, PRD1 and φX174 have been used extensively to study virus transport under various column and field conditions. Parameter estimates for each model virus were provided for the one-site model discussed above.

For the purposes of estimating log removal in saturated soil, MS2 was selected as the model virus. MS2 meets the requirements for a worst-case model virus provided the water temperature is less than 10°C (Schijven and Hassanizadeh, 2000a).

3.2 Aerosol Exposure Pathway

3.2.1 Area Source Model

An area source model was used to estimate the concentration of viruses at specific downwind distances and wind speeds. An area source model was selected over a point source due to the fact that biosolids are more typically applied over a large area rather than sprayed from a point source. The area source model takes into account the dimensions of the field where the biosolids have been or are being applied. The model equation is provided in Appendix F. The model is based on the number of particles emitted from the source per second, the mean air speed in meters per second, the diffusion factors in the x and y planes; and a microbial inactivation constant.

The rate of release from the source (Q) was estimated by back calculating the equation using sampling data. The number of viruses per m^3 of air must first be estimated in order to back calculate Q by using an aerosolized F^+ coliphage estimate of 1 PFU/m3 per 1000 PFU/g (Brooks et al., 2004). Given a concentration of viruses in biosolids, c PFU/g, the aerosolized concentration was estimated as c (PFU/g) × (1 (PFU/m^3) / 1000 (PFU/g)) = $c \times 10^{-3}$ PFU/m^3. The dispersion parameters and depth of atmospheric mixing layer are based on Pasquill's stability categories (Pasquill, 1962) and estimates from Parker et. al. (1977). There are six stability categories, A through F based on wind speed, incoming solar radiation, and cloud cover (Appendix G). Category A corresponds to hot, calm days and leads to the greatest amount of dispersion. Category D corresponds to windy days or the transition times of dawn and dusk and is the most frequently occurring stability category. Category F corresponds to nights with low winds and very little dispersion. The parameter estimates for the dispersion in the lateral and vertical direction of flow and the depth of the mixing layer were obtained from Parker et al. 1977 based on the above stability categories by wind speed. The microbial inactivation constant is based on a value obtained experimentally (Dowd et al., 2000).

3.2.2 Empirical Model

An empirically derived site-specific model based on a point source spray tanker was developed that predicts downwind concentrations of coliphage by distance (Brooks et al., 2005). The model was derived by fitting a simple linear regression of $\log_{10}$ concentrations of coliphage on downwind distance based on data from a field study. This model has the limitation that it is specific to a particular site with respect to meteorological conditions and the point source application. This model is useful for predicting downwind concentrations of viruses for the given set of conditions; however for purposes of selecting a more flexible model for this risk assessment, the above area source model was used.

3.3 Form of Output from Exposure Pathway Models

For each exposure scenario a distribution of attenuation was derived that characterizes the level of pathogen reduction during transport. The distributions were developed by varying the more uncertain parameters of each model using Monte Carlo simulation. For example, mass transfer rate data for the unsaturated groundwater model are minimal and thus the parameter estimates for the mass transfer coefficients are highly variable (Faulkner et al., 2002). By varying this parameter, in addition to other highly variable parameters, distributions for the attenuation in the unsaturated zone were developed. Many of the groundwater scenarios contain multiple types of media, e.g. scenario 5 is unsaturated soil followed by saturated soil. For these scenarios, the total log removal is the sum of the log removals from two distributions, one for the unsaturated portion and one for the saturated portion.

Parameters in the aerosol model are similarly varied to generate a distribution for this pathway.

3.3.1 Simulation Results

3.3.1.1 Unsaturated Model

Monte Carlo simulations were run for various types of soil (sand, silt, clay) using echovirus parameters. The log removal in any soil type other than sand was so large that essentially all the viruses are removed. Thus, the model used the parameter estimates for the sand with echovirus only (19 in all).

The first five parameters in Table 3-2 remained fixed. The last nine parameters in Table 3-2 were varied according to truncated normal distributions, bounded by ± 2 times their standard deviations. The five parameters shown in Table 3-3 were assumed to vary according to a multivariate normal distribution. The corresponding variance-covariance matrix is shown in Table 3-3.

Table 3-2. Unsaturated Model Parameters Estimates.

Parameter	Description	Distribution
Constants		
μ (g m^{-1} h^{-1})	Viscosity of water	4.72×10^{3}
ρ_w (kg m^{-3})	Density of water	1.00×10^{3}
σ (g h^{-2})	Surface tension of water	9.62×10^{8}
T (°C)	Temperature	1.17×10^{1}
θ_m (m m^{-3})	Moisture Content	2.00×10^{-1}
Varied Parameters		
$\log_{10}\lambda$ ($\log_{10}$ h^{-1})	Inactivation rate for suspended viruses	N(-2.406×10^{0}, 1.62×10^{-1})
$\log_{10}\lambda^*$ ($\log_{10}$ h^{-1})	Inactivation rate for soil-sorbed viruses	N(-2.684×10^{0}, 1.62×10^{-1})
κ (m h^{-1})	Suspended to solid sorbed virus mass transfer coefficient	N(1.34×10^{-3}, 1.80×10^{-3})
κ^* (m)	Suspended to air sorbed virus mass transfer coefficient	N(9.27×10^{-3}, 1.80×10^{-3})
r_v (m h^{-1})	Virus radius	N(1.38×10^{-10}, 1.00×10^{-11})
K_d (m^{-3} g^{-1})	Equilibrium distribution coefficient	N(7.44×10^{-4}, 2.53×10^{-3})
ρ (g m^{-3})	Soil bulk density	N(1.58×10^{6}, 1.42×10^{5})
r_p (m)	Mean soil particle radius	N(4.71×10^{-4}, 1.60×10^{-5})
α_z (m)	Vertical hydrodynamic dispersivity	N(5.59×10^{-3}, 1.00×10^{-4})

Table 3-3. Correlated Unsaturated Model Parameters Variance-covariance Matrix.

	θ_r	θ_s	*log(α)*	*log(n)*	*log(K_s)*
θ_r (m m^{-3})	0.00001	0.00003	-0.00009	0.00012	0.00042
θ_s (m m^{-3})	0.00003	0.00103	0.00021	-0.00038	0.00191
log(α) (log_{10} m^{-1})	-0.00009	0.00021	0.00113	-0.00185	-0.00446
log(n) (log_{10})	0.00012	-0.00038	-0.00185	0.00593	0.01506
log(K_s) (log_{10} m^{-1})	0.00042	0.00191	-0.00446	0.01506	0.04731

The results of the simulations (*n*=25,000) are provided in Table 3-4. The mean log-removal is quite large even for small barrier sizes. For a barrier of only 0.25 meters, the mean log removal was estimated as 19.2. Thus, it appears that if the barrier consists of unsaturated soil then essentially all remaining viruses are removed. However, the estimated 4th percentile is 5 log removal, and thus there is a small probability of achieving a low log removal. Figure 3-2 provides the distributions of log removal in sand for 0.25 and 0.5 meters. These results are consistent with the Virulo software results. The Virulo software predicts that the probability of failing to achieve a 19 log removal of viruses in 0.25 meters of sand is 0.47. Similarly, Virulo estimates the probabilities of failing to achieve 0.05 and 289 log removal as 0 and 0.91 respectively.

Table 3-4. Log Removal in Unsaturated Zone.

Distance (m)	Mean	SD	Min	Max
0.25	19.2	18.1	0.05	289.2
0.5	37.8	34.4	0.08	320.8
1	72.7	58.8	0.24	323.3

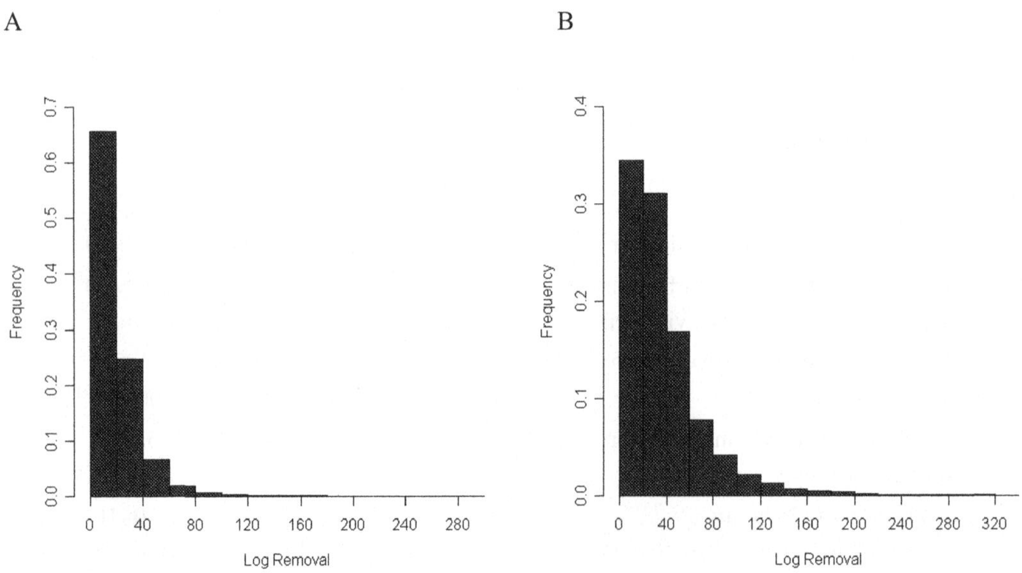

Figure 3-2. Distribution of Removal in Sand A) 0.25 meters; B) 0.5 meters.

WERF

3.3.1.2 Saturated Model

As mentioned above, the dispersivity and velocity were calculated directly from the experimental conditions and thus remained constant in the simulations. The inactivation rate was also set to a constant value in the simulations. For the remaining three parameters, the standard deviations were calculated from the point estimates and 95% confidence intervals. These parameters were assumed to follow truncated normal distributions bounded on the low end by the larger of zero and two times their standard deviations, and on the upper end by two times their standard deviations. The parameter values are provided in Table 3-5.

Table 3-5. Saturated Model Parameter Estimates.

Parameter	Description	Distribution
Constants		
α_L	Dispersivity (m)	4.3×10^{-3}
v	Velocity (m day^{-1})	3.5×10^{0}
μ_l	Inactivation rate coefficient of free phages (day^{-1})	4.5×10^{-2}
Varied Parameters		
$k_{att\,l}$	Attachment rate coefficient	$N(2.5 \times 10^{0}, 1.5 \times 10^{-1})$
$k_{det\,l}$	Detachment rate coefficient	$N(1.4 \times 10^{-1}, 3.1 \times 10^{-2})$
μ_{sl}	Inactivation rate coefficient of attached phages (day^{-1})	$N(4.8 \times 10^{-1}, 6.6 \times 10^{-2})$

The results of the simulations (n=25,000) with the above parameter values are provided in Table 3-6. The log removal is significantly lower than the unsaturated model, as the saturated layer is not as efficient in removing viruses as the unsaturated layer. Due to the smaller number of parameters and their lower variability, the variability in the estimated log removal is lower as compared to the unsaturated results (Figure 3-3).

As a result of selecting the steady-state solution to the kinetic model, the log removal is linear with respect to distance as shown in Figure 3-4. It should be noted that this model does not capture the non-linear portion of the log removal. The breakthrough curves from field experiments show an initial high log removal followed by a more linear log removal. By assuming the steady state situation, this phenomenon is not captured.

Table 3-6. Log Removal in Saturated Zone at Varying Distances.

Distance (m)	Mean	SD	Min	Max
5	1.23	0.09	0.93	1.54
15	3.68	0.27	2.74	4.62
30	7.35	0.53	5.68	9.15

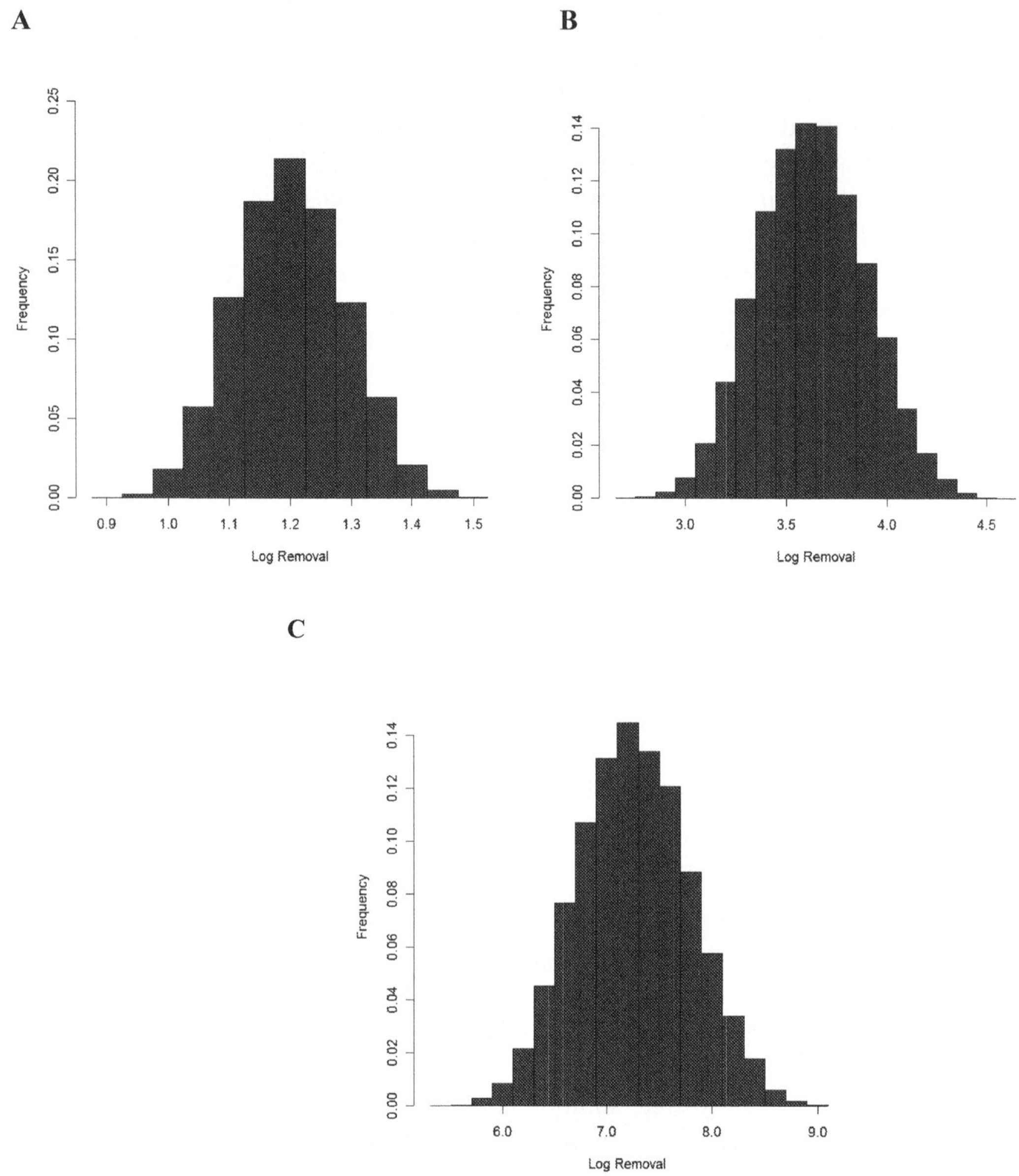

Figure 3-3. Distribution of Removal in Saturated Soil at Distance: A) 5 meters; B) 15 meters; C) 30 meters.

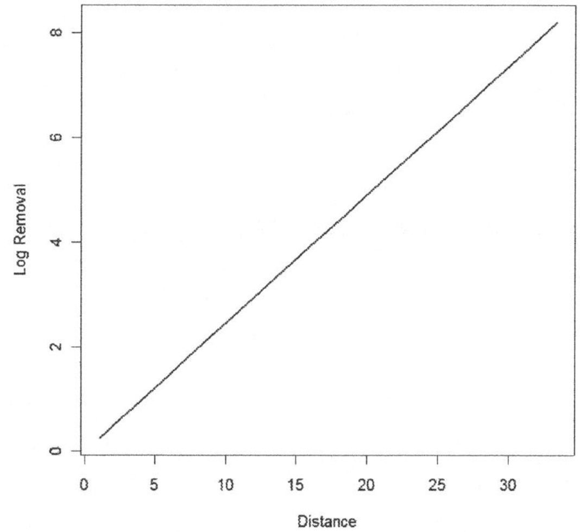

Figure 3-4. Log Removal by Distance (m) in Saturated Soil.

3.3.1.3 Aerosol Model

Simulations (n=25,000) were run for each of three downwind distances (50, 100, and 250 meters), and three wind speeds (2, 5, and 10 meters per second) for a total of nine simulations. The dimension of the plot used for the simulations was 100 meters by 100 meters (x_0 by y_0). The crosswind distance (y) and the vertical source height (σ_{Z0}) were both set to zero. Table 3-7 provides the values for the dispersion factors in the lateral and vertical directions of flow as well as the depth of the mixing layer based on stability category are provided in Table 3-7. The stability category is based on wind speed. These parameters remained fixed in the simulations.

Only 1 parameter was varied in the simulation. A mean value of 2.66×10^{-2} (Dowd et al., 2000) and a standard deviation of 10^{-2} were used as parameters of a truncated normal distribution bounded on the low end by the larger of 0 and 2 times the standard deviations, and on the upper end by two times the standard deviations.

Table 3-7. Fixed Parameters in Aerosol Model.

Stability Category	Wind speed (m/s)	Vertical (σ_E)	Lateral (σ_A)	Depth of Layer (H_m)
A-B	2	0.1745	0.3491	30
C-D	5	0.08727	0.1745	100
D-F	10	0.0524	0.0524	1000

The results of the nine simulations are provided in Table 3-8. The downwind concentration increases with wind speed and decreases with downwind distance. The minimum downwind distance was selected as 30 meters based on a survey of states setback distances (personal communication Greg Kester, HAC member). The results produce lower values than that of found in Dowd et al. (2000) due to the aerosolized concentration used to back calculate the rate of release from the source. Dowd et al. (2000)used a value of 0.05 PFU/m^3 which was found to be high based on more current data (Brooks et al., 2005). These simulations were based on an initial concentration of viruses in the biosolids of 0.2 PFU/g. The estimated aerosolized concentration was then 2×10^{-4} PFU/m^3 based on the aerosolized phage estimation of 1 PFU/ m^3 per 1000 PFU/g.

Table 3-8. Downwind Concentrations by Downwind Distance and Wind Speed.

	Downwind Distance (m)					
	30		100		250	
Wind speed (m/s)	Mean	SD	Mean	SD	Mean	SD
2	1.34×10^{-4}	1.09×10^{-5}	1.72×10^{-5}	4.61×10^{-6}	6.49×10^{-7}	4.27×10^{-7}
5	1.36×10^{-4}	2.19×10^{-6}	4.50×10^{-5}	4.85×10^{-6}	6.71×10^{-6}	1.80×10^{-6}
10	1.49×10^{-4}	4.85×10^{-6}	5.26×10^{-5}	2.81×10^{-6}	1.67×10^{-5}	2.23×10^{-6}

CHAPTER 4.0

RISK ESTIMATES

In this chapter the researchers used the risk method elaborated in Chapter 1.0 to construct models for risk assessment. The features of these models are that exposure estimates incorporate data from raw sludge monitoring, treatment processes, and post-treatment monitoring of biosolids. These exposure estimates are in turn used to estimate risk using both of the models presented in Chapter 1.0, the individual-based model, which assumes no secondary transmission or immunity, and the population-based model that accounts for secondary transmission and immunity.

The specific scenarios used in the assessments presented in this chapter are detailed in Chapter 2.0. In brief, a treatment process is specified that includes either one or two anaerobic digesters and may or may not be followed by lime treatment. The biosolids are applied to the surface of agricultural land twice per year for a period of three days for each application. Exposure can occur in a variety of ways. First, two groups may be exposed to direct ingestion. Children in residences nearby the application site are exposed during the application period and workers involved in the treatment process or application are exposed five days a week. Second, residences nearby are exposed through drinking ground water contaminated with pathogens from the biosolids, and third, residences nearby are exposed through aerosol transport. The details of the groundwater and aerosol transport models are described in Chapter 3.0.

This chapter presents the risk assessment methodology in three parts: first the data are described, including raw sludge pathogen concentrations, treated sludge pathogen concentrations, soil ingestion rates, and treatment efficacy; next the statistical approach is presented that is used to estimate pathogen concentrations when some measurements are non-detectable; and finally five risk models are presented, each subsequent one adding on features to the prior one. Of paramount importance in this chapter is the demonstration of the risk-modeling framework. The risk estimates produced in this chapter are meant only to illustrate the application of the risk model. As better or site-specific data become available or as conditions are varied, the model is a viable framework to examine the impact of these new data.

4.1 Data

4.1.1 Raw Sludge Data

Monitoring data from four utility sites were obtained (see Chapter 2.0 for details). The focus of the risk assessment analyses is based on one year of data (2003-2004) from Dataset 1 (Table 2-8). This dataset consists of a full year of monthly enteric virus data in raw sludge. The detection limit for the sampling method for this utility is 1 PFU/4 grams per dry weight.

4.1.2 Treated Sludge Data

For all of the utility monitoring sites, with the exception of site 2, the analysis of samples in the treated sludge resulted in all non-detects. This is primarily due to the fact that most utilities that have enteric virus data are producing Class A biosolids. Site 2 has a multi-step treatment process and provides concentrations in samples at each step. The analysis of samples from the final treatment step for this site resulted in all non-detects; however, there was detectable virus earlier in the treatment process.

One published data source (Gerba et al., 2002) provides monthly concentrations in Class B mesophilic anaerobically treated biosolids (Table 2-5).

For the risk assessment presented here it is assumed that all treated sludge data are non-detects.

4.1.3 Treatment Data

Table 4-1 provides a brief summary of the treatment data collected in the literature review that were used to estimate the log removal for anaerobic digestion. The first five estimates of the total $\log_{10}$ reduction for mesophilic digestion range from 0.5 to 2. The table provides one reference for the $\log_{10}$ reduction per day at three plants, which range from 0.003 to 0.03. The only thermophilic digestion data source provides an estimated $\log_{10}$ reduction of 3. The final reference provides an equation of inactivation per day and inactivation of viruses based on temperature.

Based on these data sources, an estimate of 2-log is used in the model as the mean attenuation of viruses during anaerobic digestion in each of the risk assessment models.

Table 4-1. Treatment Data Summary.

Type of Anaerobic Digestion	Log_{10} Reduction	Reference
Mesophilic	0.5-2	(Sorber and Moore, 1986b)
Mesophilic	1	(Gerba et al., 2002)
Mesophilic	1	(U.S. EPA, 1991b)
Mesophilic	1.36	(U.S. EPA, 1991b)
Mesophilic	1.05	(U.S. EPA, 1991b)
Mesophilic (3 plants)	0.03, 0.01, 0.003 (all per day)	(Goddard, 1981)
Thermophilic log reduction	3	(U.S. EPA, 1991b)
Equation for inactivation based on temperature	0.420-13.623*(incubation temp) per day	(Hurst, 1989)

4.1.4 Ingestion Rate Data

The data collected in the literature review for soil ingestion ranged from 26 - 480 mg per day (Table 2-7). The high-end estimate of 480 mg/day corresponds to an adult performing eight hours of yard work twice per week. Based on these data, the ingestion rate used in each of the risk models was 100 mg/day

4.2 Estimating Concentration Values with Censored (Nondetectable) Data

One approach for estimating or imputing non-detected values is to use a method called regression on order statistics (ROS). This procedure imputes the non-detected values within the range of 0 to the non-detect level, which for the risk assessment example presented in this report is 1 PFU/4 grams. The details of the method are presented in Shumway et al. (2002)and are shown in Figure 4-1. The general idea is that the procedure fits a line to the detected values and then uses this fit to extrapolate down below the detection limit to impute the non-detected values. The sample mean and variance are then calculated from the set of observed and imputed values.

This method assumes that the values are normally distributed. Environmental contaminant data are commonly assumed to be log-normally distributed. Based on this assumption, the ROS method is applied to the log-transformed values. The variance of the lognormal distribution of viruses ($\hat{\sigma}_y^2$) is estimated by calculating the sample variance of the set of observed and estimated values. Transforming the estimates back to the original (non log) scale often introduces transformation bias. To help correct this bias, the Quenouille-Tukey Jackknife method is used to estimate the mean of the lognormal distribution ($\hat{\mu}_y$) and variance of this mean estimate ($\hat{\sigma}_{\hat{\mu}_y}^2$) (Shumway et al., 2002). Note that the mean estimate is assumed to follow a normal distribution with parameters $\hat{\mu}_y$ and $\hat{\sigma}_{\hat{\mu}_y}^2$. The ROS method is unbiased when the assumption of the lognormal distribution of the untransformed values is correct but is not robust against distributional departures from the lognormal. Therefore it is important to check this assumption.

Figure 4-1. ROS Method for Estimated Censored Values.

Estimating the non-detect values. We have n_0=5 observations, y_i, i=1,2,…,5, each below detection limit U=1 and n_1=7 observations greater than U. Using an appropriate transformation (e.g. x_i=ln(y_i)), we assume the observations are independently and normally distributed and have common mean μ_x and variance σ_x^2, the mean and variance will satisfy the equation $x_i = \mu_x + \sigma_x \Phi^{-1}(P_i)$ where $P_i = \Pr\{X_i \le x_i\}$ and $\Phi^{-1}(\cdot)$ is the inverse of the cumulative normal distribution function. A regression is performed on the inverse order statistics. The procedure is to replace the probabilities by the adjusted ranks so that the regression equation becomes

$x_i = \mu_x + \sigma_x \Phi^{-1}\left(\frac{i-3/8}{n+1/4}\right) + \varepsilon_i$ where $i= n_0+1, n_0+2, \ldots, n_0+ n_1$, n = 12, and the estimators for μ_x and σ_x are estimated by least squares.

So for the data set used in this report, the 7 observed values are 5, 2, 1, 1, 7, 6, 31 PFU/ 4 gm (Table 2.8). The ordered statistics procedure requires that the values are ordered, where the first 5 are non-detects and the following 7 are: 1, 1, 2, 5, 6, 7, 31. These values are next transformed, x_i=ln(y_i). The normal scores for each of the 7 transformed values are estimated using the above function, $\Phi^{-1}(\cdot)$. The resulting 7 values are then used to estimate the two parameters, $\hat{\mu}_x$ and $\hat{\sigma}_x^2$, of the above linear function. The 5 non-detected values $\hat{y}_i = \exp(\hat{x}_i)$, i=1,…,5 are then estimated from the same linear model.

Estimating the Variance of the Distribution of Viruses. The variance of the distribution of viruses is estimated by the sample variance $\hat{\sigma}_y^2$ of the set of 12 observed and estimated values.

Estimating the mean. Let $\tilde{\mu}_y$ represent the sample mean of the set of observed and imputed values transformed back to the original scale. Let $\mu_{[-i]}$ represent the sample mean with the ith observation deleted, i=1,…,n.

Let $\overline{\mu}_{[-]} = n^{-1}\sum_{i=1}^{n} \mu_{[-]}$ be the sample mean of these estimators. The Quenouille-Tukey Jackknife estimator is $\hat{\mu}_y = n\tilde{\mu}_y - (n-1)\overline{\mu}_{[-]}$.

Estimating the variance of the mean. The variance of the above Quenouille-Tukey Jackknife estimator is:

$$\hat{\sigma}_{\hat{\mu}_y}^2 = \frac{n-1}{n}\sum_{i=1}^{n_1}\left(\mu_{[-i]} - \overline{\mu}_{[-]}\right)^2 .$$

4.3 Models

The ROS method described in the previous section is used to provide estimates of the non-detectable values in the raw sludge dataset. This complete dataset in used as input into our risk assessment models. The following section describes five risk models. The first is the most simple, assuming a constant treatment efficacy; the second model contains a single, 1st order, anaerobic digester; the third model contains two digesters in series; and the fourth and fifth extend the second model by adding exposure pathways and a population-level transmission model respectively.

4.3.1 Measures of Risk

Three types of risks were estimated from the models: individual-level single event risk, individual-level annual risk and population level attributable risk. All three risk measures are based on the beta-Poisson dose-response function that estimates the probability of infection given a specified dose,

$$P = 1 - (1 + d / \beta)^{-\alpha},$$

where P represents the probability that an exposed individual will become infected, d represents the dose, and β and α are parameters for a specific virus. The virus used in all risks models is rotavirus with $\beta = 0.42$ and $\alpha = 0.26$.

4.3.1.1 Individual-level Single Event Risk

This risk measure is defined as the risk associated with an individual's one time exposure to biosolids. This is estimated using the probability of infection provided by the dose-response function. The dose is estimated using one of the models described in the following sections and an ingestion rate for the given exposure pathway. For example, a risk event is described as an individual ingesting 100 mg of biosolids. The model for direct ingestion with a 1st order digester estimates a concentration in the biosolids; for illustrative purposes a concentration of 0.05 PFU/gram is used. The viral dose based on the 100 mg ingestion and the given concentration is 5×10^{-3}. The individual-level risk is in this example, 3×10^{-3}. This means that for every 1000 times an individual is exposed to this dose they will likely become infected or sick 3 times.

4.3.1.2 Annual Individual-level Risk

This risk measure is defined as the risk associated with an individual's exposure to biosolids over a yearlong period. It is calculated from the individual-level risk as

$$Annual\ Risk = 1 - (1 - P)^t$$

where P is the individual level risk and t is the time exposed in days. Continuing the example from the previous section, assuming an occupational exposure with a five-day work week, and thus 260 days of exposure per year, the annual individual-level risk is 0.55. This means that a worker exposed to these conditions every day of their work year, can expect to likely become infected or sick, 0.55 times in any given year.

4.3.1.3 Population-level Annual Attributable Risk

The cumulative incidence (CI) is defined as the number of new cases of disease during a one-year period divided by the total population at risk. For each population model scenario, the simulations were run twice, once with biosolids exposure present and once with biosolids exposure absent to obtain CI_b and CI_0. The attributable risk due to biosolids (AR) was calculated as the difference between CI_b and CI_0. For example, let's say when biosolids are present, the model produces $CI_b = 1.56$ cases per person per year, and when biosolids are absent, the model produces $CI_0 = 1.44$ cases per person per year. The attributable risk due to biosolids would be AR = 0.12 cases per person per year. In this example, the number of cases due to biosolids is a small percentage of the total cases.

The dose-response function is used to estimate the risk of a given dose of viruses. However, the risk in the population model is also based on risk from person-to-person transmission in addition to the risk associated with ingestion of viruses.

4.3.2 Model 1: Estimating Individual-level Risk for Direct Ingestion Exposure with Constant Treatment Removal

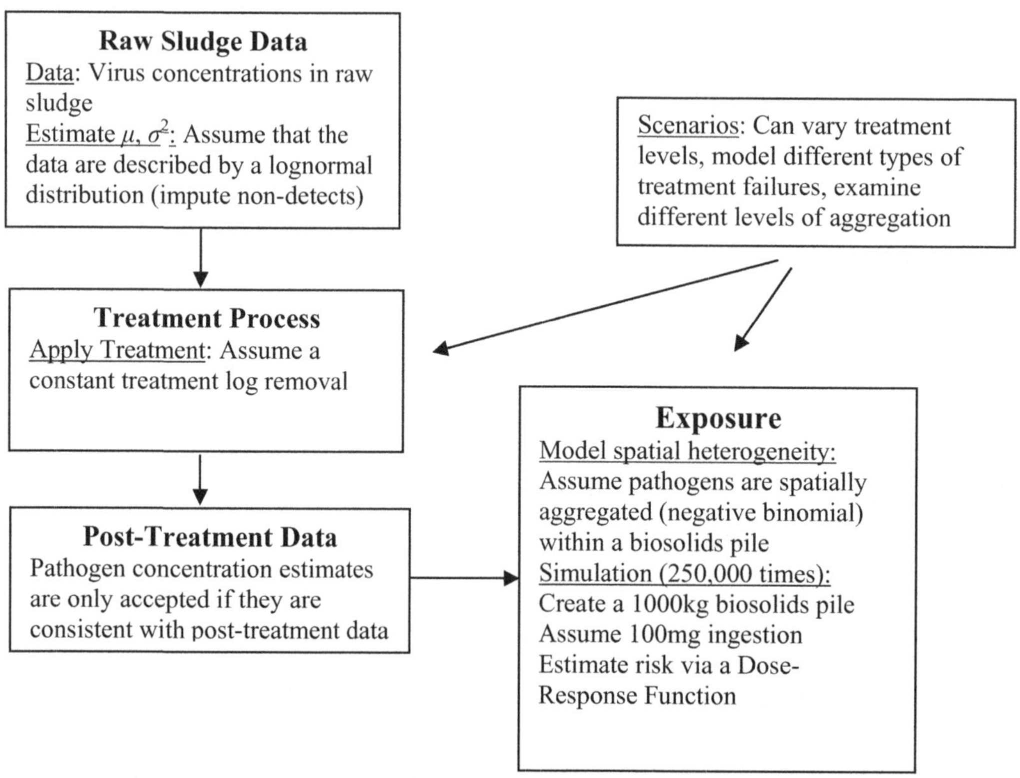

Figure 4-2. Flow Diagram for Model 1.

The model consists of four components (Figure 4-2). The first component is the input distribution that represents the concentration of viruses in the raw sludge. The variability in pathogen concentrations is represented by a log-normal distribution with parameters $\hat{\mu}_y$ and $\hat{\sigma}_y^2$.

The uncertainty is represented by describing the mean concentration as a normal distribution with parameters $\hat{\mu}_y$ and $\hat{\sigma}^2_{\hat{\mu}_y}$. The resultant distribution of concentrations in the sludge is a joint distribution of the log-normal distribution LN($\hat{\mu}_y$, $\hat{\sigma}^2_y$) where mean is normally distributed (N($\hat{\mu}_y$ $\hat{\sigma}^2_{\hat{\mu}_y}$).

The second component is treatment. The pre-treatment distribution of the concentrations is attenuated by a fixed constant treatment, e.g., a 2-log removal.

The third component of the model is a boundary condition. Post-treatment concentrations for Class A biosolids must all be below the detectable limit of 1 PFU per 4 grams. Thus, using either simulation method for Class A biosolids, a necessary condition on the post-treatment concentrations is that they are all below the detectable limit. For Class B biosolids, the condition is not as strong and there may be detectable concentrations even after treatment. So a less stringent boundary condition, therefore, is possible when estimating risks from Class B biosolids. In all cases, the boundary condition used is based on the measured post-treatment concentrations. For the purposes of demonstrating the model the measured output is always assumed to be below detectable levels. Monitoring, however, is limited in both space and time. The boundary condition, therefore, requires that concentration levels are below detectable levels 99% of the time. The value of 99% simply reflects the fact that if 1% of the samples are above the detection limit, detecting viruses in a standard monthly, single grab sample will be rare.

The fourth component is exposure. This component models the spatial heterogeneity in a biosolids pile. The pathogens are assumed to be aggregated according to a negative binomial distribution. This distribution is commonly used to model aggregated spatial distributions. A nice property of this distribution is that as the aggregation parameter k approaches ∞, the negative binomial coincides with a Poisson distribution, which has a random spatial pattern. Thus, it allows the flexibility of creating biosolids piles that are anywhere from highly aggregated (or clumped) to randomly dispersed with respect to the pathogen distribution.

Simulations were run for a variety of scenarios based on different treatment levels and different aggregation levels. Some scenarios include treatment failures, e.g., a given percentage of the biosolids pile was not treated at all. Each of these simulations is run 250,000 times, each time creating a 1000 kg biosolids pile. For this risk assessment exercise it is assumed that 0.1 g of direct ingestion occurs from each 1000 kg pile. That is, the risk scenario is a biosolids worker ingesting 0.1 g from a 1000 kg biosolids pile. A rotavirus dose-response function is used to estimate risk.

For the solution to converge, 250,000 simulations are required for each scenario. For computational efficiency, therefore, the 1000kg pile is divided into 100g pieces. The negative binomial distribution is discrete and thus the concentrations for each 100g piece are in the range 0,1,2,.... PFU per gram. With a treatment of 2-log removal, this results in a minimum possible concentration of 10^{-4} PFU per gram. Given an ingestion rate of 0.1 grams/day, this results in a minimum ingestion of 10^{-5} PFU, which gives a minimum risk of approximately 6×10^{-6} (or 6 cases of infection for every 1 million events of exposure).

The simulation process (repeated 250,000 times for each scenario) is as follows:

1) Randomly sample from normal distribution with mean $\hat{\mu}_y$ and variance $\hat{\sigma}^2_{\hat{\mu}_y}$ to obtain μ_{cp}, the mean pathogen concentration. The units of μ_{cp} are PFU per 4 grams.

2) Randomly sample from lognormal distribution with mean μ_{cp} and variance $\hat{\sigma}_y^2$ to obtain c_p, the actual pathogen concentration. The units of c_p are PFU per 4 grams. Convert the units of c_p into PFU per 1 gram to obtain c_p^*.
3) Create a 1000kg pile by randomly selecting 1000 negative binomial numbers with mean c_p^* and aggregation parameter k. The units of c_p^* are PFU per gram.
4) Apply a constant treatment over entire pile.
5) Randomly sample 0.1 gram from the 1000kg biosolids pile and calculate risk using rotavirus dose-response function.

This procedure is repeated 250,000 times producing 250,000 estimates that describe the resulting risk distribution.

4.3.3 Model 2: Estimating Individual-level Risk for Direct Ingestion Exposure Modeling a 1st Order Anaerobic Digester

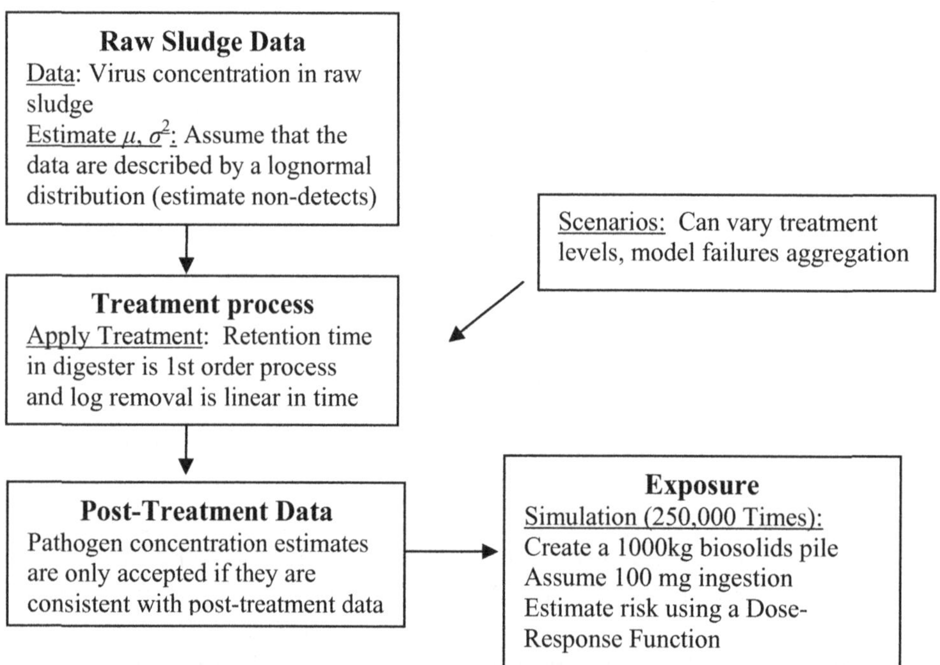

Figure 4-3. Flow Diagram for Model 2.

Model 2 differs from Model 1 in that the treatment is not constant across the whole biosolids pile. By assuming that the sludge in the digester is well mixed, the retention time in the treatment digester can be modeled as a *1st order process*. A first order process means that the retention time in the digester follows an exponential distribution; i.e., if the biosolids material is divided into discrete volumes or boluses, the retention time, and therefore the treatment efficacy, will vary from bolus to bolus. The mean log removal of the whole biosolids train is constant, but the specific log removals of a given bolus will vary as determined by an exponential distribution. The log-removal is in turn assumed to be linearly related to retention time. For example, a 3 log removal at a retention time of 15 days results in a slope of 1/5, and thus at 10 days the log removal is 2. Simulations were run for different slopes of the log removal line, as well as complete treatment failure for a proportion of the biosolids mass. The treatment failure means that a portion of the biosolids mass was not treated due to a breakdown in the treatment process. Although not done in this study, one could also define a treatment failure as variation in the treatment efficacy over time. For Model 2, simulations are run 25,000 times, each time creating a 1000kg biosolids mass. The researchers again assumed 0.1 grams of direct ingestion and use a rotavirus dose-response to estimate risk.

Model 2 converges more quickly than Model 1, and, therefore, only requires 25,000 simulations. This method, however, is computationally intensive, requiring that the 1000kg pile is divided into 1000g pieces. Since the treatment log removal is not constant in this model, there is no restriction on the minimum dose and thus the risk is not bounded as in Model 1.

The simulation process for Model 2 (repeated 25,000 times for each scenario) is as follows:

1) Randomly sample from normal distribution with mean $\hat{\mu}_y$ and variance $\hat{\sigma}^2_{\hat{\mu}_y}$ to obtain μ_{cp}, the mean pathogen concentration. The units of μ_{cp} are PFU per 4 grams.
2) Randomly sample from lognormal distribution with mean μ_{cp} and variance $\hat{\sigma}^2_y$ to obtain c_p, the actual pathogen concentration. The units of c_p are PFU per 4 grams. Convert the units of c_p into PFU per 1 gram to obtain c_p^*.
3) Randomly sample from an exponential distribution with mean retention time β_{ret} to obtain retention time in digester to obtain t_{ret}. The units of t_{ret} are days.
4) Calculate treatment removal based on line with slope l / β_{ret} where l is a log removal that varies by scenario. Log removal=$t_{ret} l / \beta_{ret}$. Calculate and retain c_p^{**}, the concentration of viruses in the 1000g piece, by attenuating c_p^* using the above calculated log removal.
5) Repeat the above 4 steps 1000 times to accumulate a 1000kg pile.
6) Randomly sample 0.1 gram from the 1000kg biosolids pile and calculate risk using rotavirus dose-response function.

This procedure is repeated 25,000 producing 25,000 estimates that describe the resulting risk distribution.

4.3.4 Model 3: Estimating Individual-level Risk for Direct Ingestion Exposure Modeling a 2nd Order Anaerobic Digester

Model 3 is the same as Model 2 except there is the addition of a second digester placed in series. To simulate this scenario, the researchers simply perform step 3 in Model 2 twice and sum the two retention times, each with a mean of 7.5 days. The total mean retention time over the two digesters is 15 days. Note: the overall time in the digesters is now a gamma distribution; i.e., it is the sum of two exponentials, $\boldsymbol{\beta}_{ret_1} + \boldsymbol{\beta}_{ret\,2}$.

4.3.5 Model 4: Estimating Individual-level Risk for Exposure From Aerosol or Groundwater Exposure Modeling a 1st Order Anaerobic Digester

4.3.5.1 Groundwater Exposure

Up to this point, the risk simulation models were based on direct ingestion from the anaerobic digester. The researchers now consider that scenario where the biosolids are applied to soil, enter the groundwater, and are subsequently ingested from contaminated well water. Viruses from the biosolids are attenuated as they pass through soil, the degree to which depends on the type of media. In the context of the previous models, this movement through soil can be thought of as additional attenuation of the viruses in the biosolids.

There are three types of groundwater media that are modeled: 1) non-porous (fractured bedrock or karst), 2) unsaturated soil and 3) saturated soil. Groundwater scenarios may include one or more of these types of media. The strictly non-porous scenarios assume a fixed level of attenuation. Fate and transport models for unsaturated and saturated soils were obtained from the literature, as were the distributions of the corresponding model parameters. Output distributions of viral concentrations are obtained using Monte Carlo simulation techniques that randomly sample each parameter distribution as well as an input viral concentration output distribution of viral concentrations is obtained. See Chapter 3.0 for more details of the model structures.

This exposure model serves to attenuate the virus concentration due to transport of biosolids through the groundwater. These three types of media combined with the many possibilities for barrier depths result in a large number of simulation scenarios. Based on previous simulations of the attenuation in unsaturated soil, a barrier depth of greater than 0.5 meters results in log removals so high as to remove essentially all remaining viruses. Thus, 0.5 meters is the maximum barrier depth used for unsaturated soil. Saturated soil is not as efficient in removing viruses, and thus the maximum barrier depth is 30 meters. The following construct for the groundwater model is intended to capture as much of the variability as possible while not creating an overly complex model.

1) Non-porous (fractured bedrock or karst) media only.
    a) Assume an attenuation of ~1-3 log removal of viruses.
2) Non-porous media followed by 5, 15 or 30 meters saturated soil (3 scenarios).
3) Unsaturated soil followed by saturated soil (6 scenarios).
    a) 0.25 meters of unsaturated soil followed by either 5, 15 or 30 meters of saturated soil
    b) 0.5 meters of unsaturated soil followed by either 5, 15 or 30 meters of saturated soil

Thus there are 10 groundwater scenarios. Each of these scenarios is run in conjunction with a 1st order anaerobic digester model.

In order to estimate a dose corresponding to the groundwater exposure, several additional assumptions were made. The first assumption is that a rainfall event occurs immediately after the 1000 kg pile of biosolids is applied to land. The second assumption is that a proportion of the viruses from the pile leach from the solids and move down through the soil or karst with the rainwater. The leaching proportion estimate used in the model is 8%. This estimate is based on a study that was aimed at investigating the leaching and transport of viruses. The study found that less than 8% of coliphage initially present in biosolids leached out of the biosolids-soil matrix (Chetochine et al., 2006). The third assumption is that the well is directly downstream of where the biosolids are applied (Figure 4-4). The attenuation in the specific medium (media) for the given scenario is calculated based the downstream distance using the models discussed in Chapter 3.0. This attenuation is applied to the remaining 8% of viruses that leached during the rainfall event. The fourth assumption is that the rainwater containing the viruses is diluted with clean water with a ratio of 1:1. The fifth assumption is that the viruses arrive over a period of three days, with a third of the viruses arriving each day. The viruses are distributed in the well-pumped water over the three days, based on an average of 150 gallons (567 liters) of water pumped per day. An ingestion rate of 1.2 liters of water per day was used. Based on this dose, the rotavirus dose-response function is used to estimate the single-event risk. The annual risk is also estimated from the model,

$$Annual\ Risk = 1-(1-r)^{d},$$

where r is the single event risk and d is the number of days exposed. To estimate the annual risk, it is assumed that the biosolids are applied 2 times per year with a 3-day exposure each time for a total of 6 days of exposure per year.

Figure 4-3 provides a schematic for the flow of the viruses after a rainfall event for the scenario of unsaturated soil followed by saturated soil. The arrows represent the flow of viruses to the well that is directly downstream of the biosolids pile.

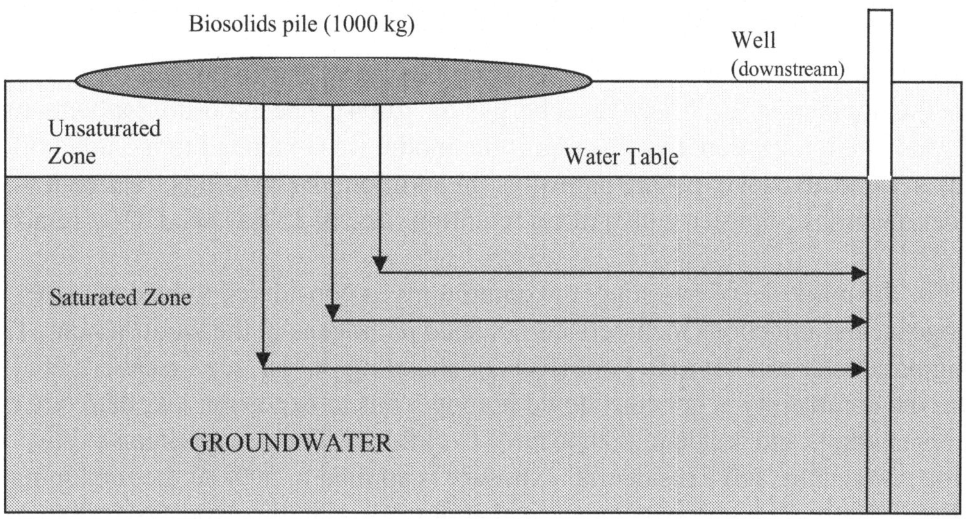

Figure 4-4. Groundwater Flow Diagram.

4.3.5.2 Aerosol Exposure

A second possible exposure pathway that is examined is the scenario in which the biosolids are applied to land and the viruses are subsequently transported via aerosolization. An area source model was used to estimate the downwind concentrations of viruses. The size of the area source plot was assumed to be 100 meters by 100 meters. A 1000 kg pile of biosolids was applied to the given plot. The exposed person breathing the air in each scenario was assumed to be directly downwind of the biosolids application site.

Based on the literature review, the aerosol scenarios include wind speeds of 2, 5, and 10 meters/second at downwind distances of the biosolids application site of 30, 100, and 250 meters. These combinations cover typical meteorological conditions as well as typical distances from biosolids application sites, which range from 30 meters to 300 meters (Personal communication Greg Kester, HAC member). The nine distinct scenarios were run in conjunction with the 1st order digester model.

In order to calculate the dose for the aerosol exposure pathway, the average human breathing rate of 0.83 m^3 per hour, and exposure duration of 8 hours was used. Thus the dose was calculated as,

$$Dose = cd \times 0.83 \times 8,$$

where *cd* is the downwind concentration in viruses per m^3. The rotavirus dose-response function was used to estimate the individual-level single event risk based on this dose. The annual risk is also estimated from the model. To estimate the annual risk, it is assumed that the biosolids are applied two times per year with a three-day exposure each time for a total of six days of exposure per year.

4.3.6 Model 5: Estimating Population-level Risk For Direct Ingestion Exposure Modeling a 1st order Anaerobic Digester

A dynamic model was developed during Phase I of this investigation and is used here to estimate the population level risk (Eisenberg et al., 2004). The exposure pathway modeled during Phase I was direct ingestion. For this report the model was extended to include both the groundwater and aerosol exposure pathways. In addition, the biosolids treatment was modeled as a 1st order anaerobic digester with a mean retention time of 15 days and a log removal of 2 at 15 days.

For the purposes of this study the community is considered to be open to external sources of pathogens, and infectious members do not shed pathogens in the environment. Thus the shedding parameter was set to 0. For effects of shedding on risk see Eisenberg et al. (2004). In addition, the community is not considered homogenous to exposure; i.e., there are occupational exposures to adults and residential exposures to children. The occupational exposure is applied to 1% of the population, and a residential exposure is applied to 50% of the remaining 99% of the population. Within each group, occupational and residential, the exposure is assumed homogenous.

For the residential exposure, the biosolids are applied two times per year with a three-day exposure each time. For the occupational exposure, workers are exposed five days per week, 52 weeks per year. The model is run for 1,500 days to reach a steady state and subsequently run for another 365 days. The number of disease onsets during the last 365 days is counted resulting in a one-year cumulative incidence. The attributable risk for biosolids (AR) is estimated as the difference in the cumulative risks when biosolids are present and when they are not (Eisenberg et al., 2004).

There are three exposure scenarios for the population-level risk model. The first focuses on occupational only exposure, the second on residential only exposure, and the third on both occupational and residential exposures within a community.

4.3.6.1 Parameter Estimates

All parameter values with the exception of shedding, secondary transmission and the pathogen concentration in the environment due to sources other than biosolids, are set to the values according to the decision tree that classified a high attributable risk, determined in Phase I of this project (Eisenberg et al., 2004). Other parameters are set to the middle value as determined in Phase 1 of this project (Eisenberg et al., 2004), including those that were either: 1) not factors in the decision tree (see Figure 4 in Eisenberg et al., 2004); and 2) not related to shedding, secondary transmission or the pathogen concentration in the environment due to sources other than biosolids. As indicated above, the shedding parameter (φ) is set to 0. The total number of pathogens in the environment (W) is equal to,

$$W = W_{\text{ext}} + \varphi ,$$

where

$$W_{\text{ext}} = V_b + V_e,$$

where V_b is the number pathogens from biosolids and V_e is the number of pathogens from sources other than biosolids. Thus, when shedding is set to 0,

$$W = V_b + V_e.$$

The estimate of V_e is determined such that when pathogens from biosolids are absent ($V_b = 0$) and secondary transmission is 0, the incidence is 20 cases per 100,000. The secondary transmission parameter (β_{s2}) is then estimated by the value that results in 40 cases per 100,000 with V_b, V_e and φ as above. This secondary transmission estimate thus doubles the risk of a composite environmental exposure that excludes biosolids. This particular estimate was not based on data; it was used in this analysis to simply illustrate the potential role of secondary transmission in risk assessment.

The rotavirus dose-response function was used to estimate the risk associated with the exposure to biosolids.

4.4 Results

4.4.1 Fitting Raw Sludge Data From the Four Monitoring Sites

The results of imputing the non-detect values for Dataset 1 using ROS are provided in Table 4-2. The five imputed values range from 0.044 to 0.605. With these imputed values, a complete dataset can be used to estimate pre-treatment concentrations. The pre-treatment concentrations are estimated using a lognormal distribution. The lognormal function requires the identification of two parameters, the sample mean and variance. The estimated mean, $\hat{\mu}_y$, and variance, $\hat{\sigma}_y^2$, are 4.53 PFU/4 grams and 75.5 respectively. The variance estimate for the lognormal distribution characterizes the variability of the data. There is also uncertainty, which is a function of the sample size (in this case n=12). Assuming that the sampling errors are normally distributed, the estimated variance of the mean concentration, $\hat{\sigma}_{\hat{\mu}_y}^2$, is 6.3.

Table 4-2. Imputed Values for Nondetects: Dataset 1.

Date	Raw Data	Imputed Values
3-Aug	**<1**	**0.044**
3-Sep	**<1**	**0.124**
3-Oct	**<1**	**0.234**
3-Nov	**<1**	**0.388**
3-Dec	**<1**	**0.605**
4-Jan	5	5
4-Feb	2	2
4-Mar	1	1
4-Apr	1	1
4-May	7	7
4-Jun	6	6
4-Jul	31	31

These parameter estimates were calculated for all four data sources described in Chapter 2.0 (Table 4-3). Dataset 2: Site 2, Dataset 3 and Dataset 4 did not contain any non-detects in the data and thus the ROS techniques were not required. For these data sources, $\hat{\mu}_y$ was estimated by the sample mean, $\hat{\sigma}_y^2$ was estimated by the sample variance and the Quenouille-Tukey Jackknife was used to estimate $\hat{\sigma}_{\hat{\mu}_y}^2$. The mean concentration estimates varies by site from 4.5 to 30.3 PFU/4 grams.

The results presented in the remaining sections are all based on Dataset 1.

Table 4-3. Estimated Concentrations for 4 Monitoring Datasets.

Data Source	$\hat{\mu}_y$ (PFU/ 4 g)	$\hat{\sigma}_y^2$	$\hat{\sigma}_{\hat{\mu}_y}^2$
Dataset 1	4.5	75.5	6.3
Dataset 2: Site 1	3.8	25.3	1.2
Dataset 2: Site 2*	3.7	19.9	0.8
Dataset 3*	43.4	926.3	132.3
Dataset 4*	30.3	914.8	76.2

*No non-detects

4.4.2 Model Simulation Results

The first two steps in simulating Models 1 and 2 are identical. These steps provide an estimate of the mean concentration in sludge (c_p^*). Figure 4-5a shows the lognormal fit for the raw sludge data (the bars represent the actual data). Figure 4-5B shows the uncertainty in the mean estimate of the lognormal distribution in Figure 4-5A. A left truncated normal distribution was used to describe the mean concentration to avoid obtaining any negative values for the mean estimate. The resulting composite distribution of pre-treatment pathogens is highly skewed (Figure 4-4C). This skewness is due to extreme values in the distribution. The maximum value is 517, as compared to the mean of 4.6, and there are 72 (0.03%) values greater than 140.

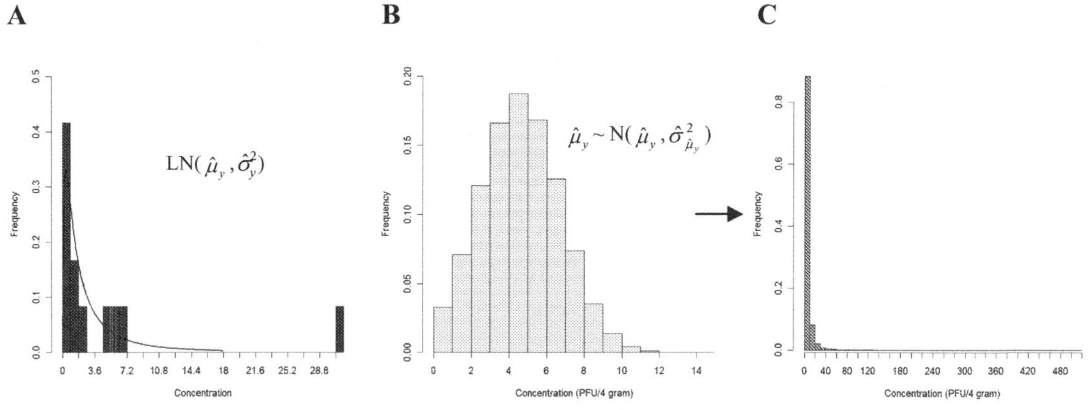

Figure 4-5. Pre-treatment Concentration of Viruses. A) Lognormal distribution of pre-treatment concentrations with imputed values, where solid line is the fitted curve and the bars are the data. B) Normal distribution of the mean value, $\hat{\mu}_y$, of the pre-treatment concentrations. This distribution is left truncated to prevent negative values. C) The composite pre-treatment concentrations based on A and B.

4.4.2.1 Model 1: Estimating Individual-level Risk For Direct Ingestion Exposure With Constant Treatment Removal

Figure 4-6 shows the spatial distribution of pathogens resulting from one iteration for the scenario with a constant treatment of 2-log removal for two different values of the aggregation parameter *k*. Each pixel in the graphs represents 100 grams of biosolids. The integer value within a pixel represents the number of pathogens present. The grid in Figure 4-5 contains 10,000 pixels (100 by 100) representing a 1000kg pile. The pathogen values for each pixel are obtained by randomly selecting from a negative binomial distribution. A sequence of 10,000 numbers is thus generated and plotted on the 100 by 100 grid. For example, the first value is plotted in position (1, 1), the second in position (1, 2), and so on. The numerical value plotted represents the concentration in that 100 g pixel. The two plots shown in Figure 4-5 have the same total number of pathogens. The difference between Figure 4-5A and 4-5B is that when the aggregation parameter is low the pathogens are more clumped together, and when it is high they are more random. When aggregated (*k*=0.001) there are more 100 g pixels with no pathogens at all, however the 100 g pixels with pathogens generally have high concentrations. In contrast, in Figure 4-6B where the pathogens are randomly distributed (*k*=2), most of the 100 g pixels contain some pathogens, but with generally lower concentrations. The mean concentration across all grids is 4×10^{-2} PFU/4 grams which is effectively the mean pretreatment concentration attenuated by 2-logs.

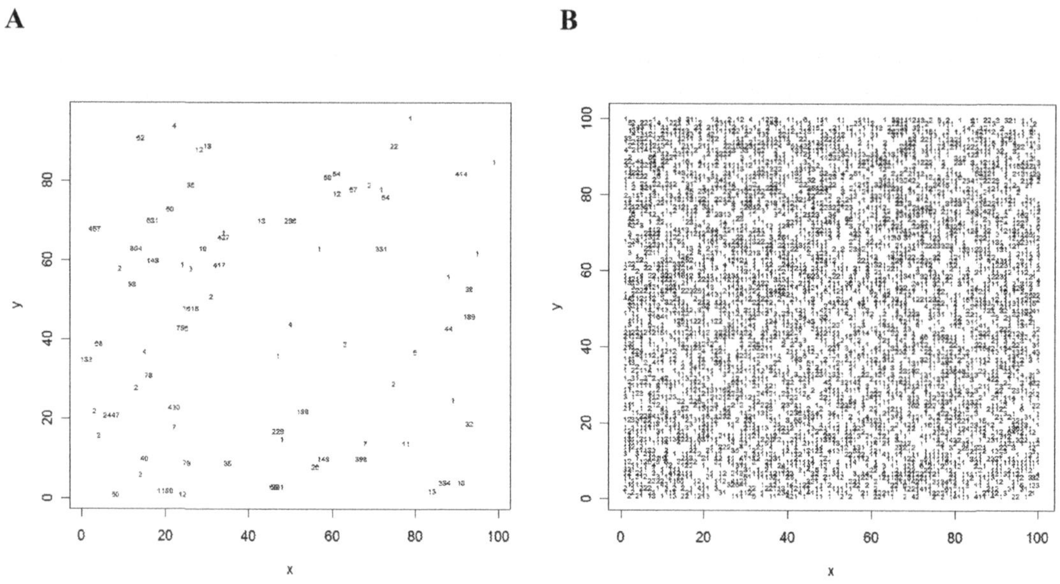

Figure 4-6. Spatial Distribution of Pathogens in a 1000 kg Biosoild Pile A) k = 0.001; B) k = 2.

The corresponding risk distributions, based on an ingestion rate of 100 mg/day, are shown in Figure 4-7. Note that for computational efficiency, the biosolids piles were created in 100 g pixels. With the biosolids created in this manner, the minimum non-zero concentration possible is 1 virus in each 100g pixel, and so a 100 mg sample has a minimum of 10^{-3} viruses. With a 2-log removal, the number of viruses in a treated 100 mg sample is 10^{-5}. Thus, the corresponding minimum non-zero risk using the rotavirus dose-response function based on the 10^{-5} viruses ingested is 6×10^{-5} (6 cases in 100,000 exposure events).

As the distribution of concentrations approaches a random spatial pattern, that is when $k = 2$, the mean risk estimates are higher as compared with the aggregated case (7×10^{-4} vs. 3×10^{-4}). The lower risk in the aggregated case is due to the large proportion of samples in the aggregated pile with no pathogens at all. This is shown in Figure 4-7A for k=0.001, where 98.9% of the samples have risk of 0 corresponding to a pathogen concentration of 0. However, the mean risk of 3×10^{-4} is still achieved as 0.1% of the samples have a risk of 0.1 or greater. Alternatively, when k=2, only 0.004% of the samples have a risk of 0.1 or greater. Thus, when k=0.001, even though the mean risk is lower than for k=2, more large concentrations and thus large risks are observed. That is, if large clumps of viruses occur in a few areas throughout the pile, the chance of ingesting them is low as most of the pile is virus free. However, if a portion of the highly concentrated clump is ingested, the risk of infection is high. The variability in the risk estimates is very similar. In these examples, risk is defined as a single event risk, which is the probability of infection or sickness in an individual exposed to virus in biosolids, as described earlier. The histograms shown below in Figure 4-7 illustrate the uncertainty and variability of that risk. For example in Figure 4-7B, there is a 23% likelihood that the single event risk is between 10^{-4} and 10^{-5}, where 10^{-4} means that 1 in 10,000 exposures will result in infection or sickness, and 10^{-5} means that 1 in 100,000 exposures will result in infection or sickness.

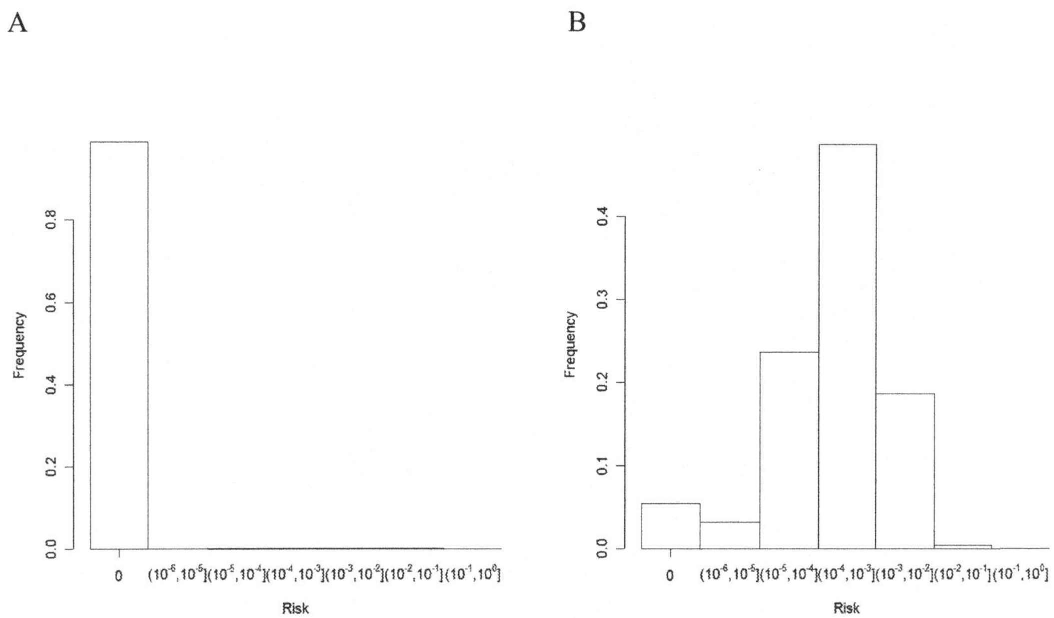

Figure 4-7. Distribution of Risk Using Model 1. A) k = 0.001, mean = 3 x 10^{-4}, σ = 8 x 10^{-3} B) k = 2, mean = 7 x 10^{-4}, σ = 2 x 10^{-3}.

Using the same simulation parameters as above but now introducing failure into the treatment process with a 1/3 probability of failure and 2/3 probability of no failure risk distributions were calculated (Figure 4-8). The choice of a 1/3 failure is an illustrative example and does not necessarily represent a realistic scenario. The idea is to show that the model is flexible enough to include failure in the treatment process and the subsequent effect on risk. The estimated mean concentration increases significantly more for the randomly distributed simulation (from 3×10^{-4} to 2×10^{-2}) than for the aggregated sample (from 7×10^{-4} to 9×10^{-4}). This is a result of the fact that even though the non-zero risks are higher, there is still a large proportion of 0 risks occurring. This contrasts from the random pile ($k = 2$) where most concentrations are non-zero.

A B

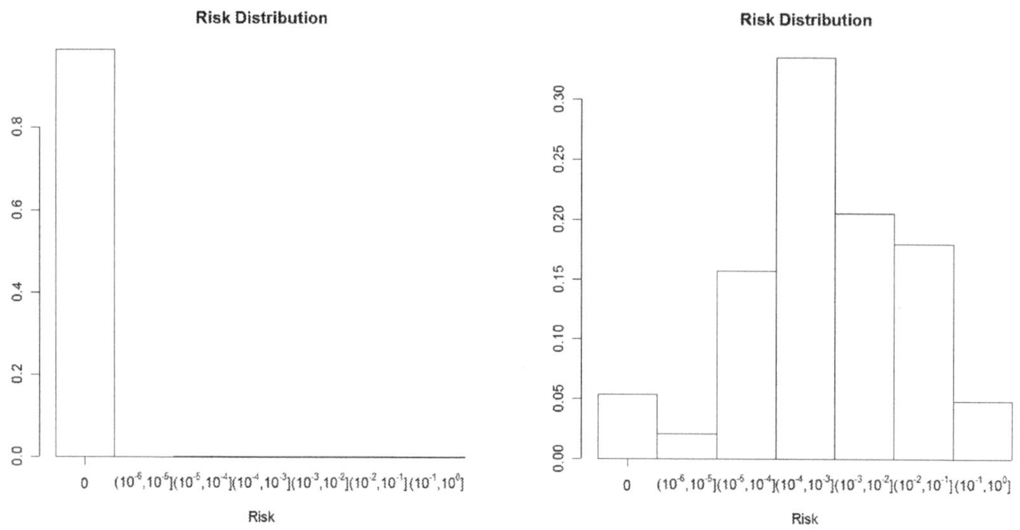

Figure 4-8. Distribution of Risk Using Model 1 Given a Failure in the Treatment Process that Occurred One Third of the Time. A) k = 0.001, mean = 9×10^{-4}, $\sigma = 2 \times 10^{-2}$ B) k = 2, mean = 2×10^{-2}, $\sigma = 4 \times 10^{-2}$.

4.4.2.2 Model 2: Estimating Individual-level Risk for Direct Ingestion Exposure Modeling a 1st Order Anaerobic Digester

Concentration estimates using Model 2 with a mean retention time of 15 days and a treatment log removal of 2 at 15 days are provided in Figure 4-9. These results are based on one iteration of the 25,000 simulations, thus the distributions are for one 1000kg simulated biosolids pile. The distribution of the retention time in the digester is exponential with a mean of 15 days. In the spatial distribution plot (Figure 4-9C), each pixel is 100 grams and the 100 by 100 grid represents the 1000kg pile. As each 100 g sample is simulated, its value is plotted sequentially. The concentration of the first 100 gram sample is plotted in position (1, 1), the second in position (1, 2) and so on. The numerical value plotted represents the concentration in that 100 gram unit. The spatial distribution of pathogens using Model 2 (Figure 4-9C) is similar to the random (k=2) case for Model 1 (Figure 4-6C). The corresponding risk distribution is provided in Figure 4-10. The mean risk estimate is almost an order of magnitude higher compared to results from Model 1 with the corresponding scenario and k=2 ($\mu_r = 1 \times 10^{-2}$ vs. 3×10^{-3}). In Model 2, the level of treatment is not constant as it is in Model 1 and thus some pieces of the biosolids pile receive very little treatment whereas some receive a large amount of treatment. For this reason, large concentrations that do not receive treatment skew the mean risk higher relative to Model 1. This also explains the higher variability in the risk estimates for Model 2 than that of Model 1, with coefficients of variation (cv), the ratio of the standard deviation to the mean, of 3 and 0.5 respectively.

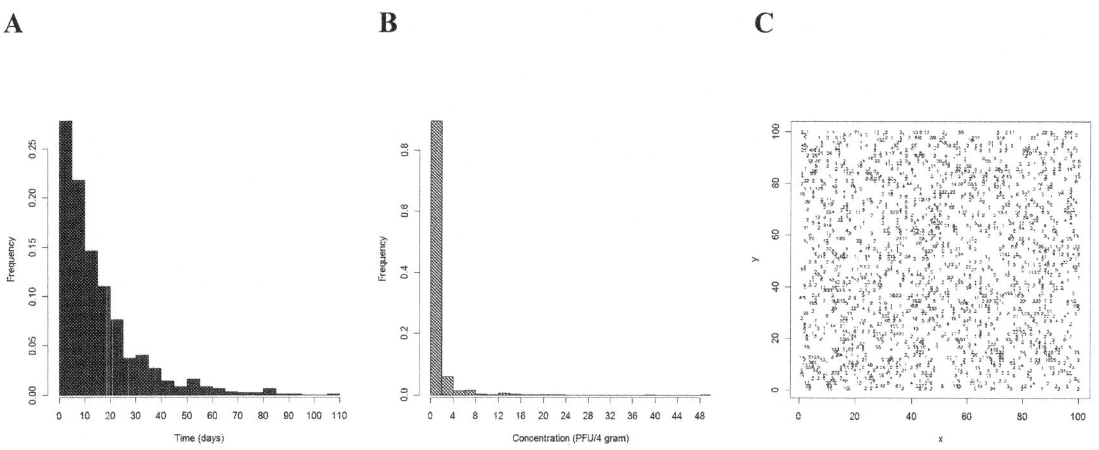

Figure 4-9. A) Distribution of Time in Digester; B) Post-treatment Distribution in Concentration; C) Spatial Distribution of Pathogen (pixel size = 100 g).

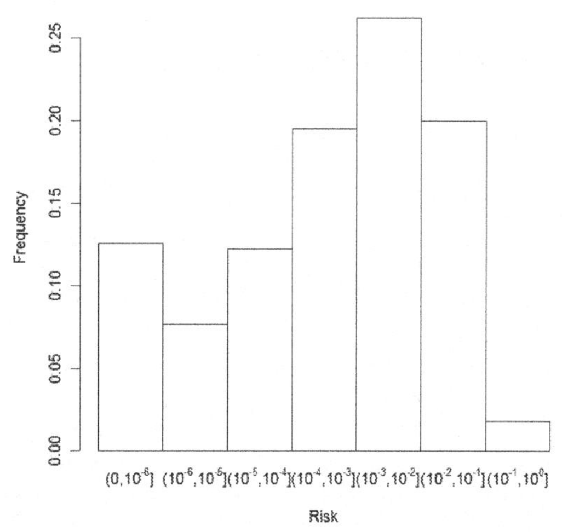

Figure 4-10. Risk Distribution For Model 2 (one digester). Mean = 1 x 10^{-2}, σ = 3 x 10^{-2}.

4.4.2.3 Model 3: Estimating Individual-level Risk for Direct Ingestion Exposure Modeling a 2nd Order Anaerobic Digester

Model 3 is identical to Model 2 except that there are two digesters in sequence as opposed to one. The mean retention time in each of the two digesters is 7.5 days to allow a comparison to the one-digester scenario with a mean retention time of 15 days. The risk distribution is shown in Figure 4-11. Compared to one digester, the mean risk is lower for two digesters ($\mu_r = 1 \times 10^{-2}$ vs. 2×10^{-3}). Similarly, it is less likely that there are long retention times as compared to the one-digester model, and therefore, the variability in the risk estimates for the two-digester model is lower than that of one-digester model ($cv = 1.1$ vs. $cv = 0.07$). Table 4-4 provides a comparison of risk estimates between the 1 and 2 digester treatment models. Most notable is the probability of a risk greater than 10^{-4}, provided in the last row of Table 4-4. This probability is twice as high for the 1 digester model (0.68) as compared to the 2 digester (0.34).

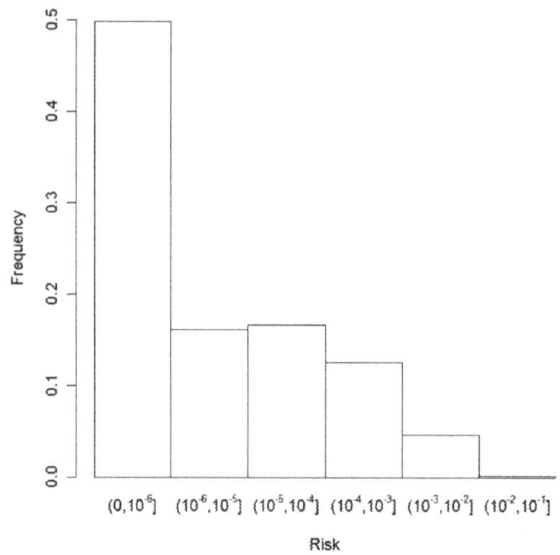

Figure 4-11. Risk Distribution For Model 3 (2 digesters). Mean = 2 x 10^{-3}, σ = 9 x 10^{-3}

Table 4-4. Comparison of Risk For 1 and 2 Digester Treatments.

Risk	Model 2 (1 digester)	Model 3 (2 digesters)
Mean	**$1x10^{-2}$**	**$2x10^{-3}$**
Median	**$8x10^{-4}$**	**$1x10^{-5}$**
75%	**$8x10^{-3}$**	**$3x10^{-4}$**
Pr(Risk>10^{-4})	**0.68**	**0.34**

4.4.2.4 Boundary Conditions

For each model, a boundary condition on the proportion of detectable samples in the simulations was set to 1%. The mean percent of detectable samples for each of the 1000 kg simulated piles for each model is provided in Table 4-5. All Model 1 simulations, with the exception of the scenario with a 2-log removal for treatment and a 1/3 probability of treatment failure, had a mean proportion of detectable samples below the boundary of 1%. For Model 2, the mean proportion of detects in the 1000kg piles over the 25,000 simulations was 4.2%. Model 3 was able to achieve the boundary condition constraint for a log removal of four or greater.

Table 4-5. Proportion of Detectable Samples For Each Model.

	K	Log Removal					
		2	**3**	**4**	**5**	**8**	**10**
Model 1 (no failure)	0.001	0.25%	0.08%				
	2	0.17%	0%				
Model 1 (failure)	0.001	0.31%	0.13%				
	2	4%	0.05%				
Model 2 (at 15 days)		17.04%	12.11%		7.60%		4.16%
Model 3 (at 7.5 day)		3.29%	1.7%	0.96%		0.28%	

The high mean proportion of detectable samples in for Model 2 is well above the boundary condition of 1%. The range of proportions over the 25,000 simulations for a log removal of 2 at 15 days is 12.6 - 22.4%. The standard deviation of the percent of detects is 1.18. Thus, all piles have a proportion of detects well above the boundary condition. As previously discussed, Model 2 does not have a constant log-removal and thus, the probability that the 1 kg pieces of the biosolids pile receive little treatment is greater than zero. This results in a higher proportion of detectable concentrations in the 1 g samples. The site from which Dataset 1 was obtained uses an anaerobic digester followed by lime treatment. An additional 1.1 log removal by lime treatment is required to reduce the mean proportion of detectable samples for Model 2 at 2 log removal to below the boundary level of 1% (Table 4-6).

Model 3 requires a log removal of 4 in order to have a mean proportion of detectable samples below the boundary level of 1%. The corresponding additional lime treatment log removal required to reduce the mean proportion of detectable samples below 1% is provided in Table 4-6.

Table 4-6. Additional Lime Treatment Required to Meet Boundary Condition.

	Log Removal					
	2	3	4	5	8	10
Model 2 (at 15 days)	1.1	1		0.85		0.65
Model 3 (at 7.5 day)	0.5	0.3				

Revised risk estimates for Models 2 and 3 when the additional lime treatment is added are shown in Figures 4-12A and 4-12B. With the additional lime treatment, for Model 2 the risk decreases from 1×10^{-2} to 1×10^{-3}. For Model 3, the risk decreases from 2×10^{-3} to 7×10^{-4}. The probability of a high risk ($> 10^{-4}$) is reduced to 0.46 and 0.16 for Models 2 and 3 respectively.

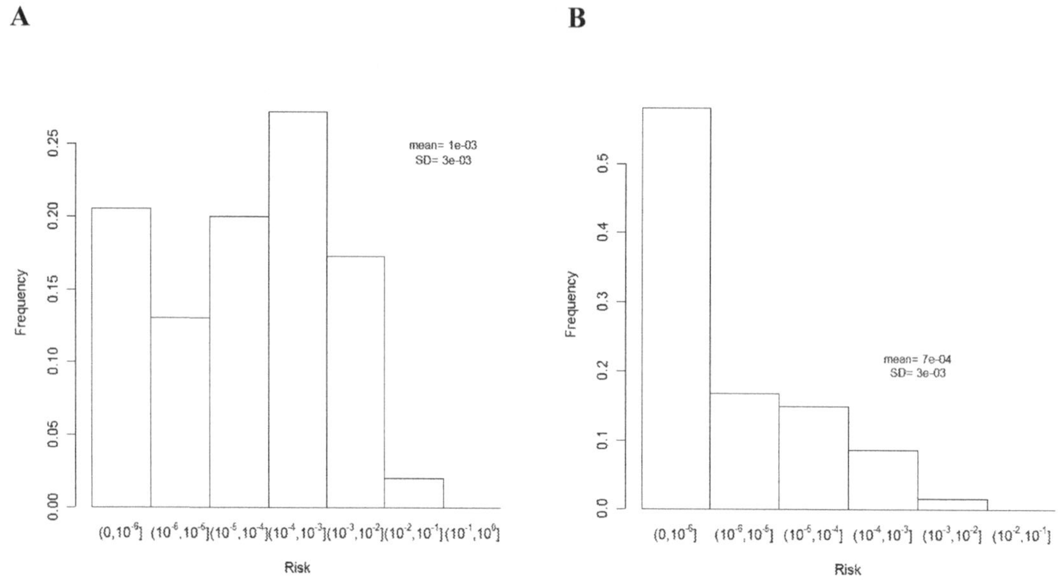

Figure 4-12. Distribution of RIsk With Additional Lime Treatment. A) Model 2 with 1.1 log removal from lime treatment, mean = 1 x 10^{-3}, σ = 3 x 10^{-3} B) Model 3 with 0.5 log removal from lime treatment, mean = 7 x 10^{-4}, σ = 3 x 10^{-3}.

4.4.2.5 Model 4: Estimating Individual-level Risk for Exposure from Aerosol or Groundwater Exposure Modeling a 1st Order Anaerobic Digester

4.4.2.5.1 Groundwater

The groundwater exposure simulations were run for every combination of the 9 groundwater scenarios with the one digester treatment model. The additional 1.1 log removal corresponding to lime treatment was also included. For each simulation, the mean retention time in the digester was set to 15 days with a corresponding log removal of 2 at 15 days. The risk distributions for each of the nine scenarios are provided in Appendix H.

The worst-case groundwater scenario is the non-porous media scenario. This scenario is essentially a direct conduit to the well, and therefore provides little attenuation relative to other media. Under this exposure scenario, the mean risk was 2×10^{-2}. This risk is higher than the direct ingestion with lime treatment which had a mean risk of 1×10^{-3}. This is due to the fact that the direct ingestion is based on a 100 mg sample of the 1000 kg biosolids pile. Thus, only a small number of viruses are actually ingested. However, with the non-porous groundwater scenario, even though there is additional attenuation, leaching and dilution, many more viruses reach the well. Thus, the number of viruses ingested is actually higher than the direct ingestion scenario. When the non-porous media was followed by saturated soil, the attenuation increased. The mean risk for this scenario was reduced to 2×10^{-3}, 2×10^{-5} and 5×10^{-8} for 5, 15, and 30 meters of saturated soil respectively.

For the scenarios that include unsaturated soil followed by saturated soil, as the barrier depths increase, the mean risk decreased to as low as 6×10^{-9} for the scenario of 0.5 meters of unsaturated soil followed by 30 meters of saturated soil.

A summary of the risks associated with these various models as a function of the saturated zone barrier depths is provided in Figure 4-13. Figure 4-13A provides the mean risk against the saturated barrier depth, which is preceded by non-porous media. The pattern is nonlinear; i.e., the decrease in mean risk is larger for small barrier depths compared with large barrier depths. Figures 4-13B and C provide the mean risk as a function of saturated barrier depth when preceded by unsaturated soil of depths 0.25 and 0.5 meters respectively. A similar nonlinear pattern in mean risk is observed for both of these scenarios.

Thus, for the groundwater exposure, the mean risks range from 2×10^{-2} to 6×10^{-9} for the worst and best case scenarios respectively. This corresponds to a single event risk. The corresponding estimated annual risks based on 3 days exposure for biosolids applied two times in the year ranges from 1×10^{-1} to 4×10^{-8}.

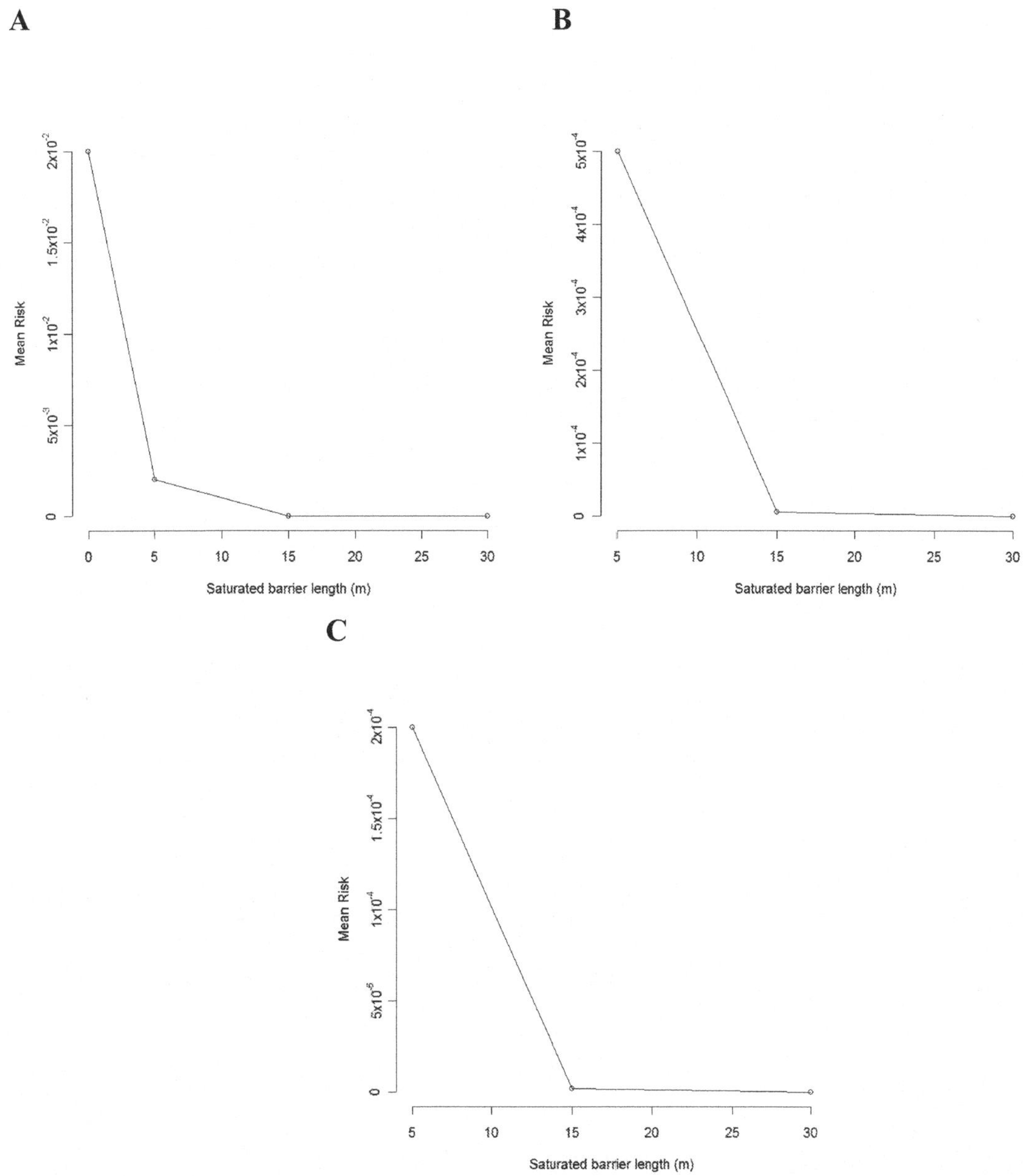

Figure 4-13. **Mean Risk Against Barrier Depth (Groundwater Scenario).** A) Non-porous media followed by saturated soil, B) Unsaturated soil = 0.25 m followed by saturated soil, C) Unsaturated soil = 0.5 m followed by saturated soil.

4.4.2.5.2 Aerosol

The aerosol exposure simulations were run for each of the nine aerosol scenarios with the 1st order digester treatment model with the additional lime treatment. The results for only the 30 and 250 meter downwind distances are included in this section. The results of the 100 meter downwind distance are provided in Appendix I. For each simulation, the mean retention time in the digester was set to 15 days with a corresponding log removal of 2 at 15 days. The risk distributions for each of the nine scenarios are provided in Appendix I.

In general, as the downwind distance from the source increases, the attenuation increases and thus the mean risk decreases. Similarly, as the wind speed increases, the attenuation decreases and therefore the mean risk increases. For a downwind distance from the source of 30 meters, the wind speed does not have any effect on the mean risk. After rounding the risk estimates, there is no observable difference for the three wind speeds. This is due to the fact that the exposure site is quite close to the source and thus the effect of wind speed is small as compared to larger downwind distances. Figure 4-13A summarizes this relationship.

Increasing the distance from 30 meters to 250 meters has a more noticeable impact on the mean risk as it ranges from 6×10^{-6} at a wind speed of 2 meters per second to as high 5×10^{-5} at 10 meters per second. Figure 4-14B shows that the increase in mean risk with wind speed is nearly linear. Thus, at longer downwind distances, the wind speed has a larger effect on attenuation and thus risk.

Based on a individual mean risk of 7×10^{-5}, assuming that biosolids are applied twice per year with a 3-day, 8-hour exposure each time, the corresponding mean annual risk is 7×10^{-4}. Similarly, the lowest risk scenario at 250 meters downwind at 2 meters per second wind speed corresponds to a mean annual risk of 3×10^{-5}.

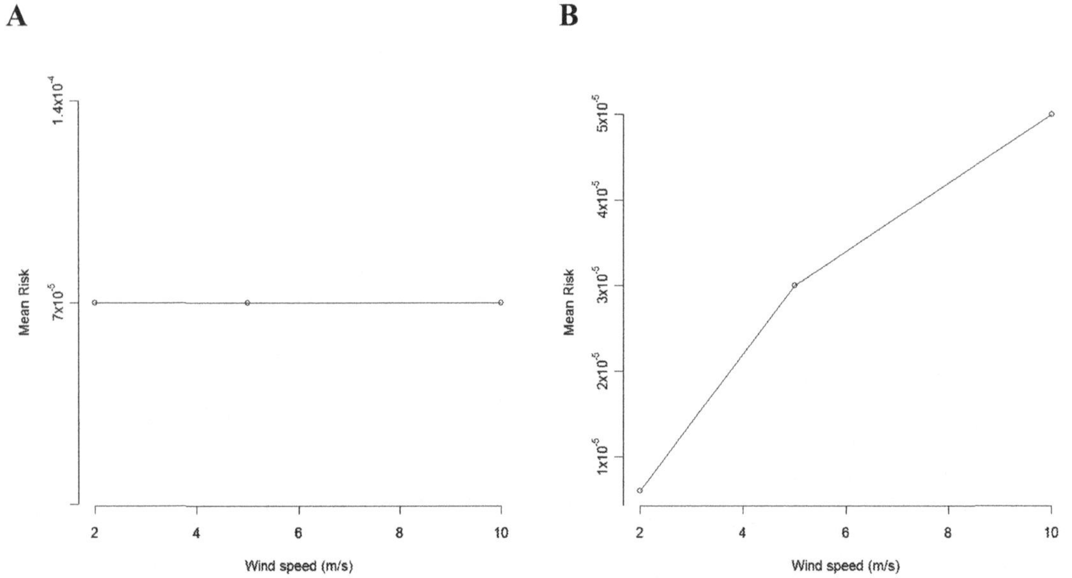

Figure 4-14. Mean Risk Against Wind Speed (Aerosol Scenario).
A) Downwind distance = 30 meters, B) Downwind distance = 250 meters.

4.4.2.6 Model 5: Estimating Population-level Risk for Direct Ingestion Exposure Modeling a 1st Order Anaerobic Digester

The parameter estimates for background levels of pathogens and the secondary transmission parameter were estimated using the method outlined in Section 4.3.6.1 to achieve a background incidence level of 40 cases per 100,000 people per year. This procedure results in the background level estimate $Ve = 1.47 \times 10^{-6}$ PFU per liter and secondary transmission estimate $\beta_{s2} = 0.04$.

Each of the three scenarios was simulated with 10,000 iterations each. For the first scenario with occupational exposure only, the simulation was run with and without biosolids present to estimate the attributable risk for biosolids. Figure 4-15A provides the attributable risk distribution for this scenario. The mean annual attributable risk was 8×10^{-5}. For comparison, using model 1 with lime treatment, the individual level single event risk was 1×10^{-3}, which corresponds to an annual risk of 2×10^{-4} assuming that the occupational risk applies to 0.1% of the population with 260 days of exposure per year. The main difference between the annual risk using model 1 and the annual attributable risk using model 5 is that model 5 includes secondary transmission. This comparison, therefore, suggests that secondary transmission attenuated the risk from 2×10^{-4} to 8×10^{-5}, a reduction of 60%. In general, when simulations are run for increasing values of secondary transmission, from the baseline of 0.04 to 1.0, the mean annual attributable risk due to biosolids decreases (Figure 4-16). That is, secondary transmission becomes protective. This phenomenon was also observed in Phase 1 of this project and is due to the fact that increases in secondary transmission results in increases in infection rates due to the other environmental pathways.

The second scenario with residential exposure was again run with and without biosolids present to estimate the attributable risk for biosolids. The residential scenario applies to children only. In this scenario the proportion of children was set to 50% of the population. Figure 4-15B provides the attributable risk distribution which has a mean annual attributable risk of 3×10^{-3}.

The third population level scenario combines both the residential exposure from scenario 1 and the occupational exposure from scenario 2. Figure 4-15C provides the attributable risk distribution, which has a mean annual attributable risk of 3×10^{-3}. The majority of the attributable risk for biosolids, therefore, arises from the residential population (3×10^{-3} vs. 8×10^{-5}). This is due to the fact that the occupational population is a very small percentage of the whole population. With secondary transmission set to zero, the attributable risk is 2×10^{-3} (vs. 3×10^{-3}, a reduction of 66%).

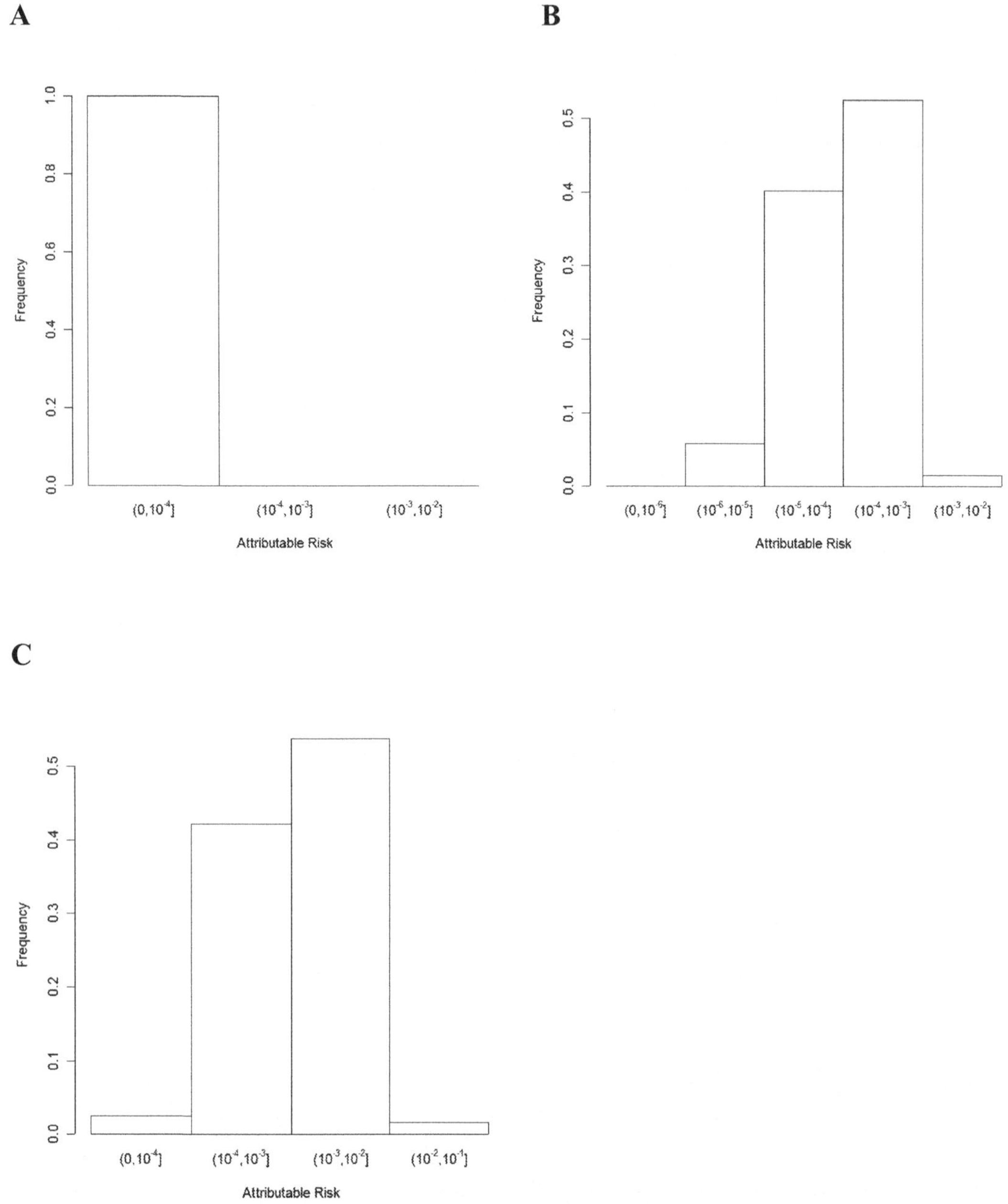

Figure 4-15 Population Model Attributable Risk From Biosolids.
A) occupational only exposure, mean = 8 x 10^{-5}, σ = 2 x 10^{-5}, B) residential only exposure, mean = 3 x 10^{-3}, σ = 4 x 10^{-3}, C) occupational and residential exposure, mean = 3 x 10^{-3}, σ = 4 x 10^{-3}.

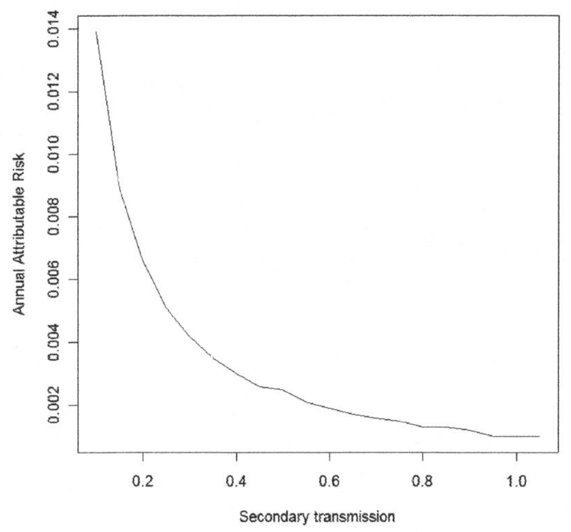

Figure 4-16. Annual Attributable Risk From Biosolids as a Function of Secondary Transmission.

CHAPTER 5.0

SUMMARY

Microbial risk assessment is a structured approach to estimating risks associated with environmental exposure to pathogens by using available environmental data and practical assumptions. In this report three sources of environmental data were used to estimate risk: pathogen monitoring of raw sludge, efficacy of treatment processes, and pathogen monitoring of post-treatment biosolids. This risk assessment approach is able to estimate risks even for Class A biosolids where post-treatment monitoring data are all below the detectable level. Using this approach risk estimates were provided for a number of different models to gain insight into risks associated with different aspects of the biosolids treatment and application process. Even though analytical data is limited, the methodology herein provides a reliable model for performing risk assessment. As more data become available, the model outputs will likewise become more robust. This project demonstrates that comprehensive risk assessments for biosolids exposure are practical. Risk estimates are provided to illustrate how the risk modeling approach presented in this report can be used both in a regulatory context to develop risk-based rules, and in an operational context to examine the benefits of changing treatment processes in the context of current application practices or proposed new practices. The following sections provide an overview of the results obtained from Models 1 thru 5. Figure 5-1 summarizes these results.

Risks Presented as Distribution

The first three models focus on how variability and uncertainty affects risk estimates, and how the design of treatment processes can minimize the impact of variability. By presenting risk estimates as a distribution rather than a point estimate, the variability and uncertainty of a risk estimate become transparent. There are a variety of sources of variability. The lognormal distribution describing the concentration of pathogens in the raw sludge (Figure 4-5A) represents the variability in concentration levels due to a variety of environmental and wastewater treatment process factors. The uncertainty due to the sample size of the monitoring data is represented by the truncated normal distribution of the mean pathogen concentration (Figure 4-5B). The composite distribution that combines these two is shown in Figure 4-5C.

5.1 Model 1

A second source of variability comes from the treatment process. In Model 1, this variability is examined by describing the spatial distribution of pathogens present in the biosolids as both a random process and as an aggregated process. The random process resulted in almost twice the mean risk than the aggregated process (7×10^{-4} vs. 3×10^{-4}), but much less variability, as measured by the coefficient of variation (cv = 0.29 vs. 27). These differences are due to the fact that in the random case there is a higher probability of ingesting pathogens than in the aggregated case, since much of the biosolids pile in the aggregated case has no pathogens. If pathogens are ingested in the aggregated case, however, the dose is much higher, and therefore so is the risk, compared with the random case.

5.2 Models 2 and 3

Whereas Model 1 accounts for the variability due to the treatment process statistically, Models 2 and 3 account for the variability by explicitly modeling the treatment process. Assuming that the digesters are well-mixed processes, they can be modeled as a first-order process; i.e., at any given time any bolus within the digester is equally likely to leave as biosolids, independent of its residence time. For Model 2 the variability ($cv = 3.0$) was less than the aggregated case in Model 1 ($cv = 27$) but more than the random case ($cv = 0.29$). A comparison of Model 2 and 3 illustrate the benefits of adding a second digester: 1) lowering the mean risk by one fifth (1×10^{-2} vs. 2×10^{-3}); and 2) decreasing the probability of there being a risk > 10^{-4} (0.68 vs. 0.34).

Using reasonable treatment efficacy values for the digesters resulted in risk levels that were not consistent with the non-detectable levels measured in the biosolids, requiring additional treatment. Enough lime treatment was applied so that 99% of the samples of biosolids were below the detectable level. This resulted in lower risks for both the one and two digester models (1×10^{-3} and 7×10^{-4} respectively).

5.3 Model 4—Groundwater

Model 4 focused on risk estimates associated with two exposure pathways, groundwater and aerosol. The risks associated with the groundwater pathways were in general lower than the direct ingestion scenarios. The exception being the non-porous model that predicted a mean risk of 2×10^{-2}, compared with 2×10^{-3} for the analogous direct ingestion estimate. Adding saturated barriers lowered the risk from 2×10^{-3} to 2×10^{-7} depending on the barrier depth. Unsaturated barrier added even more attenuation resulting in as low as a 6×10^{-9} risk depending on the barrier depth.

5.4 Model 4—Aerosol

Aerosol exposures risks were a function of the downwind distance of the exposure site to the source, and the wind speed. Risks ranged from 7×10^{-5} when the source is 30 meters from the exposure site and 5×10^{-5} when the source is 250 meters with a wind speed of 10 meters per second, to 6×10^{-6} when the source is 250 meters with a wind speed of 2 meters per second.

5.5 Risk Measures

Risk estimates for Models 1, 2, and 3 were all single-event risks. For both the groundwater and aerosol models, in addition to single event risks annual risks were estimated for residential exposure, based on 3-day exposure per application for biosolids applied two times per year. Single-event risks provide information on the probability of infection or disease of an individual that is exposed to biosolids. The annual risk estimates take into account the amount of biosolids exposure during a given year as well as the target population that is exposed. Therefore, the annual risks can provide information on the health cost to the population. Although the annual risks were only calculated for Model 4, they can be used as the risk measure for Models 1, 2, and 3.

5.6 Model 5

The annual risk estimates assume that the disease outcome of one individual is independent of the disease outcome of other individuals, as discussed in Chapter 1.0. Therefore, it is still an individual-level analysis that is scaled up to provide population estimates. When accounting for secondary transmission, a full-fledged population analysis is required, which is the focus of Model 5. Using Model 5 requires a population measure of risk; therefore, single-event risk estimates are not appropriate. Rather annual risks are estimated using an attributable risk metric, the number of cases of infection or disease that is attributable to exposure to biosolids (see Chapter 4.0 for more details of the attributable risk). For these simulations, therefore, the residential scenario, as with model 4, assumes a three-day exposure per application for biosolids applied two times per year. Occupational exposures assume workers are exposed five days per week, 52 weeks per year. Comparing the residential and occupational exposures, residential exposures dominated the overall risk estimate because the target population was so much larger, even though the yearly exposure in the occupational setting was larger. Given the scope of this work, the comparison of these two exposure scenarios is not meant to imply the importance of one exposure over the other; rather it was meant to demonstrate that the role of a risk assessment is to provide estimates for a variety of scenarios so that risk management can use this information for decision-making.

5.7 Transparency of Risk Estimates

In addition to risk estimates, a risk assessment must also provide all the assumptions used to make the risk estimates to the risk manager. Assumptions can be divided into those that come from the model structure and those that come from the parameter estimates. Chapter 4.0 contained explicit information on the assumptions used when constructing the model. For example, it was assumed that the digesters are well-mixed first order processes. Explicit assumptions were made about the groundwater and aerosol transport, as well as how the biosolids are applied, which are specified in Chapters 2.0 and 3.0. Parameter estimates also relied on data from the literature, such as the ingestion rate and treatment efficacy. These assumptions are detailed in Chapters 2.0 and 4.0. The assumptions made in these risk assessments are meant to balance between realism and simplicity. A site-specific risk assessment would need to collect as much site-specific information as possible in order to conduct a realistic risk assessment.

5.8 Conclusions

This report provides an approach to conducting risk assessments that takes advantage of pathogen data from raw sludge and data on treatment process efficacy as additional data sources to the post-treatment data. In this context the raw sludge and treatment data are considered prior information that can inform the risk estimate, and the post treatment data are used to inform the likelihood. A Bayesian structure like this one can be used to take full advantage of the available data in order to obtain the best risk estimate. As with any Bayesian analysis, if the data informing the likelihood are sparse, the posterior (or risk estimate) is strongly influenced by the prior. As the variance of the data decreases, the posterior (or risk estimate) is influenced less by the prior. Risk estimates were provided to illustrate how this risk modeling approach can be used both in a regulatory context to make risk-based rules, and in an operational context to examine the benefits of changing treatment processes in the context of current application practices or proposed new practices.

Biosolids producers can use the results of this research to interpret the results and potential health implications of biosolids monitoring data. Treatment plant engineers can evaluate the potential risk-based benefits of making operational changes to existing treatment and/or adding additional treatment processes, and environmental managers can use this research to evaluate potential capital improvements and/or land application site placement issues from a health-based perspective. Regulators at the U.S. EPA can use this framework to develop risk-based regulations parallel to the chemical contaminant rules. The risk assessment framework presented in this report provides a mechanism to discuss biosolids management microbial risk using a common metric for comparison of treatment methods, management alternatives, and potentially to set risk-based standards for microbial contaminants in biosolids. For all purposes, this report also identifies data gaps and the need for additional information to be used as input variables which will add robustness to the model.

Table 5-1. Risk Summary. Risks were defined as: Single event, the probability of disease given one time exposure; annual risk, the probability of a disease over a yearlong period; and attributable risk, the number of illnesses in a population associated with a given biosolids exposure scenario (See Section 4.3 for details). Models are in Chapter 4.0: Model 1 comprises constant treatment with a pathogen distribution in biosolids that is either random or aggregated; Model 2 comprises one anaerobic digester with or without lime treatment; Model 3 comprises two anaerobic digesters with or without lime treatment; Model 4 (groundwater) comprises Model 2 followed by transport to groundwater. Groundwater transport is through one of the seven soil types: 1) Non-porus (NP) media; 2) NP followed by 5 meters of saturated soil (5mS); 3) NP followed by 15 meters of saturated soil (15mS); 4) NP followed by 30 meters of saturated soil (30mS); 5) 0.5 meters of unsaturated soil (0.5mUS) followed by 5mS; 6) 0.5 meters of unsaturated soil (0.5mUS) followed by 15mS; 7) 0.5 meters of unsaturated soil (0.5mUS) followed by 30mS; Model 4 (aerosol) comprises Model 2 followed by aerosolized transport. Aerosolized transport is via one of three conditions: 1) exposure site 30 m downwind and wind speed of 2 m/s; 2) exposure site 250 m downwind and wind speed of 2 m/s; 3) exposure site 250 m downwind and wind speed of 10 m/s.; Model 5 comprises a population-level model using either a residential, occupational exposure scenario, or both.

	Single event risks				Annual Risks	Attributable Risks
	Mean . (μ)	Standard Deviation . (σ)	Coefficient of Variation (cv)	P(risk > 10^{-4})	(μ)	(μ)
Model 1						
Random	7×10^{-4}	2×10^{-3}	2.9	0.01		
Aggregated	3×10^{-4}	8×10^{-3}	26.7	0.68		
Model 2						
Without lime	1×10^{-2}	3×10^{-2}	3.0	0.68		
With lime	1×10^{-3}	3×10^{-3}	3.0	0.46	2×10^{-4}	
Model 3						
Without lime	2×10^{-3}	9×10^{-3}	4.5	0.34		
With lime	7×10^{-4}	3×10^{-3}	4.3	0.26	2×10^{-4}	
Model 4 (GW)						
NP	2×10^{-2}	2×10^{-2}	1.0	1	1×10^{-1}	
NP – 5mS	2×10^{-3}	2×10^{-3}	1.0	0.89	1×10^{-2}	
NP – 15mS	2×10^{-5}	4×10^{-5}	2.0	0.05	1×10^{-4}	
NP – 30mS	5×10^{-8}	2×10^{-7}	4.0	0	3×10^{-7}	
0.5mUS – 5mS	2×10^{-4}	2×10^{-3}	10.0	0.03	1×10^{-3}	
0.5mUS – 15mS	2×10^{-6}	4×10^{-5}	20.0	0.004	1×10^{-5}	
0.5mUS – 30m S	6×10^{-9}	2×10^{-7}	33.3	0	4×10^{-8}	
Model 4 (Aerosol)						
30m - 2m/s	7×10^{-5}	9×10^{-6}	0.1	0.002	7×10^{-4}	
250m - 2m/s	6×10^{-6}	4×10^{-6}	0.7	0	3×10^{-5}	
250m – 10m/s	5×10^{-5}	2×10^{-5}	0.4	0.007	4×10^{-4}	
Model 5						
Occupational						8×10^{-5}
Residential						3×10^{-3}
Occupational and Residential						3×10^{-3}

APPENDIX A

VIRUS EXPOSURE REFERENCES

Exposure Component	Reference	Pathogen	Summary
Monitoring and/or Surveillance Treatment Articles			
Biosolids Application	U.S. EPA, 1988	Virus	Review of field and lab data from various sources.
Biosolids Treatment	Gerba, et al., 2002.	Virus	A risk analysis was performed to assess which emerging pathogens would be most likely to survive Class B treatment. mesophilic anaerobically digested Class B biosolids
Biosolids Treatment	Hamparian et al., 1985	Virus	Anaerobic and aerobically digested biosolids
Biosolids Treatment	Soares, et al., 1994.	Virus	Occurrence of pathogens before and after biosolids treatment evaluated for 14 months, mesophilic anaerobically digested Class B biosolids
Fate and Transport Treatment Articles			
Biosolids: Storage	Ahmed & Sorensen, 1995	Virus	Kinetics of pathogen destruction during storage of digested and dewatered biosolids.
Biosolids Treatment	Goddard et al., 1981	Virus	Treatment comparison study on raw & digested sludge: Anaerobic mesophilic digestion and aerobic thermophilic digestion
Biosolids Treatment	Hurst, 1989	Virus	
Biosolids: Storage	Sanin, 1994	Virus	Pathogen reduction when sludge has been frozen then thawed
Biosolids Treatment	Sorber & Moore, 1986	Virus	Review of information on the survival and transport of pathogens from municipal wastewater sludge applied to the land through amending soils.

Biosolids Treatment	Ward & Ashley, 1977	Virus	Raw sludge containing 5% solids was seeded and dewatered at 21°C.

Fate and Transport Application Articles

Application: Survival in Soil	Bagdasaryan, 1964	Virus	Ten grams of soil in 100 ml. sterile flasks were infected with 1 ml. virus solution.
Application: Survival in Soil	Bitton et al., 1984	Virus	Soil cores of Eustis fine sand treated with lagooned sludge (2/3 anaerobic & 1/3 aerobic sludge).
Application: Survival in Soil	Burge and Enkiri, 1978	Virus	Virus suspensions applied to 5 soil types; no biosolids in experimental design. Compared 5 soil types as to type of adsorption isotherm and tested rates of adsorption for the soils for correlation with soil characteristics.
Application: Survival in Soil	Chattopadhyay and Puls, 1999a	Virus	Fate and transport of viruses in soils and water with respect to soprtion of viruses on soil.
Application: Survival in Soil	Chattopadhyay and Puls, 1999b	Virus	Theoretical study: thermodynamics of bacteriophage sorption on clays
Application: Survival in Soil	Chattopadhyay and Puls, 2002.	Virus	Effect of different surfactants and organic matter on bacteriophages
Application: Survival in Soil	Gerba et al., 1975	Virus	Virus and Bacteria
Application: Survival in Soil	Goyal and Gerba, 1979	Virus	Adsorption to soils (Flushing Meadows FM soil)
Application: Survival in Soil	Mawdsley, et al., 1995	Virus	Pathogens found in livestock waste and their transport thru soils.
Application: Survival in Soil	Meschke and Sobsey, 2003.	Virus	Mobility and comparative reduction of Norwalk virus with viruses and indicators through miniature soil columns
Application: Survival in Soil	Nasser, et al., 2002.	Virus	Viral reduction and inactivity in soil saturated with secondary effluent, effect of soil microbial activity on virus survival
Application: Survival in Soil	Reddy et al., 1981	Virus	Fecal coliforms, fecal stretococci, *Salmonella*, *Shigella*, viruses
Application: Survival in Soil	Sobsey et al., 1980	Virus	Settled sewage in various soil types
Application: Survival in Soil	Tierney et al., 1977	Virus	Effluent irrigated crops and virus in drained vegetable pots
Application: Survival in Soil	Zhuang, J. and Jin Y., 2003a and b	Virus	Evaluates how natural organic matter influence transport of viruses in sand and soil

Aerosol Exposure Articles

Exposure: aerosol route	Brooks et al., 2005	Virus	Empirically derived transport model used to predict downwind concentrations of viruses
Exposure: aerosol route	Brooks et al., 2004	Virus	Review article of fate and generation of bioaerosols
Exposure: aerosol route	Dowd et al., 2000.	Virus	Field study carried out to generate microbial release rates from biosolids. Data were input to bioaerosol transport model and input to MRA model, Anaerobically digested biosolids
Exposure: aerosol route	Pillai and Ricke, 2002	Virus	Bioaerosol issues associated with municipal and animal wastes.
Exposure: aerosol route	Pillai, 2002	Virus	General information on bioaerosols and public health
Exposure: aerosol route	Pillai, 2002	Virus	General information on bioaerosols and public health
Exposure: aerosol route	Teltsch and Katzenelson, 1977	Bacteria	Aerosolized bacterial concentrations from disinfected wastewater agricultural spraying.

Groundwater Exposure Articles

Exposure: groundwater pathway	Allwood, 1993	Virus	Survival data of viral pathogens and indicators in chlorinated water
Exposure: groundwater pathway	Bhattacharjee et al., 2002.	Virus	Two dimensional model for virus transport in heterogeneous subsurface porous media (groundwater aquifers)
Exposure: groundwater pathway	Deborde et al., 1999.	Virus	Field scale seeding to characterize groundwater transport of viruses, short term virus die-off
Exposure: groundwater pathway	Dowd and Pillai, 1997	Virus	Survival and transport of bacterial pathogens and indicator viruses in sandy aquifer (lab scale)
Exposure: groundwater pathway	Faulkner et al., 2002	Virus	Predictive model for fate and transport of viruses in unsaturated zone
Exposure: groundwater pathway	Faulkner et al., 2003.	Virus	Predictive model for fate and transport of viruses in unsaturated zone
Exposure: groundwater pathway	Lance & Gerba, 1984	Virus	Compared saturated and unsaturated flow under leaching conditions similar to land treatment
Exposure: groundwater pathway	Redman et al., 2001.	Virus	Filtration and deposition profiles in saturated quartz sand
Exposure: groundwater pathway	Schijven and Hassanizadeh, 2000b	Virus	In depth overview of modeling in saturated zone
Exposure: groundwater pathway	Schijven and Simunek 2002.	Virus	Reanalysis of field scale virus transport to evaluate modeling approaches, sorption models, and soil properties
Exposure: groundwater pathway	Schijven et. al., 2002	Virus	Model to estimate attenuation of viruses in saturated soil
Exposure: groundwater pathway	Schijven, et al., 2000.	Virus	Attachment and detachment rate coefficients determined for viruses in lab experimentts

Exposure: groundwater pathway	Schijven et al., 2003	Virus	Laboratory study to determine if pathogenic viruses are removed as much as MS2 through sand
Exposure: groundwater pathway	Sim and Chrysikopoulos, 2000	Virus	Model to estimate attenuation of viruses in saturated soil
Exposure: groundwater pathway	Zhuang and Jin, 2003b	Virus	Sand column experiments evaluating effects of Al-oxide on transport of viruses

APPENDIX B

VIRUS INACTIVATION RATES IN GROUNDWATER

Pathogen	Temperature (ºC)	Other condition	Inactivation rate coefficient μ (day-1)	Reference (see Pedley et al., 2006 for references)
Virus				
Coxsackievirus A9	10	Sterile	0.019	Matthess et al. (1988)
	10		0.027	
	10	Deionized	0.031	
Coxsackievirus B1	10	Sterile	0.012	Matthess et al. (1988)
	10		0.019	
	10	Deionized	0.04	
Coxsackievirus B3	3-15		0.49	Keswick et al. (1982)
Coxsackievirus B4	5		0.079	Schijven et al. (2003)
Coxsackievirus B5	16	1.2 mg/l O2	0.12	Jansons et al. (1989a)
	19.4		0.12	Jansons et al. (1989b)
Echovirus 1	12		0.24	Yates et al. (1985)
	13		0.25	
	17		0.28	
	18		0.35	
	23		0.94	
Echovirus 6	22	0.2 mg/l O2	0.25	Jansons et al. (1989a)
Echovirus 7	10	Sterile	0.032	Matthess et al. (1988)
	10		0.019	
	10	Deionized	0.038	
Echovirus 11	16	2.3 mg/l O2	0.23	Jansons et al. (1989a)
Echovirus 24	16	1.6 mg/l O2	0.12	Jansons et al. (1989a)
Hepatitis A virus (HAV)	10		0.1	Nasser et al. (1993)
	20		0.41	
	23	Filtered bottled mineral water	0.038	Biziagos et al. (1988)
	25	Sterile	0.082	Sobsey et al. (1986)
	25		0.33	
	30		0.054	Nasser et al. (1993)
Poliovirus 1	3-15		0.48	Keswick et al. (1982)
	4		0.16	Meschke (2001)
	5		0.16	Schijven et al. (2003)
	10	Sterile	0.01	Matthess et al. (1988)
	10		0.013	
	10	Deionized	0.032	
	10		0.025	Nasser et al. (1993)
	12		0.18	Yates et al. (1985)
	13		0.2	
	14	70 weeks	0.16	Meschke (2001)
	16	5.4 mg/l O2	0.21	Jansons et al. (1989a)
	16	0.2 mg/l O2	0.069	
	17		0.19	Yates et al. (1985)
	18		0.43	
	20		0.038	Nasser et al. (1993)
	22	0.06 mg/l O2	0.16	Jansons et al. (1989a)

	22		0.1	Bitton et al. (1983)
	23		0.17	Blanc and Nasser (1996)
	23		1.2	Yates et al. (1985)
	23	Filtered bottled mineral water	0.044	Biziagos et al. (1988)
	24		0.046	Bitton et al. (1983)
	25	4 weeks	0.11	Meschke (2001)
	30		0.12	Nasser et al. (1993)
	23	Filtered spring, Auvergne France	0.04439	Biziagos (1983)*
	10		0.0253	Nasser and Oman (1999)*
	34		0.046	Bitton et al. (1983)*
Rotavirus	20		0.36	Pancorbo et al. (1987)
Simian Rotavirus	3-15		0.83	Keswick et al. (1982)
	23		0.275	Gerba (undated)*
Bacteriophage				
ΦX174	5		0.012	Schijven et al. (2002b)
F-specific RNA bacteriophages	10		0.025	Nasser and Oman (1999)
	20		0.0077	Nasser et al. (1993)
	30		0.031	Nasser et al. (1993)
	10		0.0253	Nasser and Oman (1999)*
	20		0.23	Grondin (1987)*
MS2	2-5		0.03	Schijven et al. (1999)
	4		0.037	Meschke (2001)
	4		0.063	Yates et al. (1985)
	5		0.064	Schijven et al. (1999)
	5		0.082	Schijven et al. (2002b)
	7		0.0058-0.1	Yahya et al. (1993)
	12	Oxic	0.1	Schijven et al. (2000)
	12	Anoxic	0.024	
	12		0.16	Yates et al. (1985)
	12		0.065	Yates (1992)
	13		0.22	Yates et al. (1985)
	14	70 weeks	0.45	Meschke (2001)
	17		0.17	Yates et al. (1985)
	18		0.19	
	23		0.36	Blanc and Nasser (1996)
	23		0.58-1.3	Yahya et al. (1993)
	23		0.73	Yates et al. (1985)
	25	4 weeks	0.41	Meschke (2001)
PRD 1	5		0.0094	Schijven et al. (1999)
	5		0.044	Schijven et al. (2002b)
	7		0.01-0.1	Yahya et al. (1993)
	12		0.054	Schijven et al. (2000)
	23		0.035	Blanc and Nasser (1996)
	23		0.12-0.3	Yahya et al. (1993)
	23		0.23	Yahya et al. (1993)*

*Indicates this estimate was obtained from a personal communication from Phil Berger

APPENDIX C

SETBACK REQUIREMENTS BY STATE

STATE	SPECIFIC KARST PROHIBITION	DEPTH TO GW (ft)	DEPTH TO BEDROCK (ft)	SW SETBACK (ft)	SETBACK TO SINKHOLES OR DIRECT CONDUITS (ft)	MAX SOIL PERMEABILITY (in/hr)	COMMENTS
AK		10		50-100			
AL	NO	2	2	500	300	NA	SEPTAGE
AR							
AZ							
CA							
CO							
CT							
DE	NO	1.67	1.67	50	NA	NA	
FL	NO	2	NA	100 - 1000	200	NA	
GA							
HI		4		50			
IA		NA		35			
ID							
IL	NO	1.5	1.5 - 3*	200	NA	6 - 20*	DRAFT RULES
IN		1.67		33 - 300			
KS	NO	NA	NA	33	NA	NA	
KY	200 - 300	NA	4	200-300	200-300	6	
LA	NATURAL STABLE SOILS	3	STABLE SOIL	100-1000	CASE BY CASE	IF > 3 CBC	
MA		3		100			
MD	NO	2	2	25-100	NA	NA	
ME		1.25		33-300			
MI	NO	2.5	NA	50-150	50-150	NA	
MN	NO	3 TO 5	3 TO 5	50-200	50-200	NOT DIRECTLY	
MO							
MS		Case Specific		250-2640			
MT							
NC		3		50-100			
ND		5		33			
NE							
NH	NO	2 to 4	4-Feb	75-125	NA	NA	
NJ	MANY RESTRICTIONS	2	2	33	CASE BY CASE	NA	
NM							
NV	NO	3	NA	33	NA	NA	
NY		2		200			
OH	NO	NA	3	33	100 - 300	NA	
OK	NO	2	NA	100	NA	NA	
OR		2 to 4		50-200			
PA	NO	3.3	NA	33 - 100	100	NA	
RI		2		200			
SC							
SD							
TN		NA		100-200			
TX		2 to 4		33-200			
UT		NA		33			

VA	YES	1.5	1.5	25-50	25	CBC
VT	NO	3	3	50-100	NA	NA
WA		Case Specific		33-200		
WI	NO	3	3	100-200	100-200	6
WV		2		50		
WY						

APPENDIX D

EQUATIONS FOR THE FATE AND TRANSPORT OF VIRUSES IN UNSATURATED SOILS

Virulo calculates the attenuation factor *A* as:

$$A = e^{\Gamma_2(\gamma)L}$$

A is interpreted as for example, if *A*=0.0001 then $-\log_{10}(0.0001) = 4$ meaning a 4-log attenuation or 99.99% attenuation of active viruses.

The components of A are calculated as follows:

$$\Gamma_2(\gamma) = \frac{\overline{V} - \sqrt{\overline{V}^2 + 4D_z\gamma}}{2D_z}$$

1. $\overline{V} = q/\theta_m$

1.1 $q = K(\theta_m)$

1.1.1 $K(\theta_m) = K_s S_e^{1/2}\left[1-\left(1-S_e^{\frac{n}{n-1}}\right)^{1-1/n}\right]^2$

1.1.1.1. $S_e = \dfrac{\theta_m - \theta_r}{\theta_s - \theta_r}$

2. $D_z = \alpha_z \dfrac{q}{\theta_m} + \mathcal{D}_e$

2.1. $\mathcal{D}_e = \mathcal{D}/\tau$

2.1.1 $\tau = \begin{cases} \dfrac{\theta_s^{\,2}}{\theta_m^{\,11/5}} & \text{if } \theta_m \le 2 \\ \dfrac{\theta_s^{\,2}}{\theta_m^{\,7/3}} & \text{otherwise} \end{cases}$

2.1.2 $\mathcal{D} = \dfrac{k_b T}{6\pi\mu r_v}$

3. $\gamma = \lambda + \dfrac{\lambda\rho}{\left(\dfrac{\theta_m}{K_d} + \dfrac{\lambda^*\rho}{k}\right)} + k^\circ$

3.1. $k = \kappa a_T$

1.3.1.1. $a_T = \dfrac{3(1-\theta_s)}{r_p}$

3.2. $k^\circ = \kappa^\circ a_T^\circ$

$$3.1.2.\ a_T^\circ = \frac{\rho_w g \theta_m}{\alpha\sigma}\left[\left[S_e^{-1}\right]^{\frac{1}{1-1/n}} - 1\right]^{1/n}$$

The parameters are defined as:
q is the specific discharge
θ_m is the moisture content
K_s is the saturated hydraulic conductivity
n is a water retention curve fitting parameter
θ_r is the residual soil water content.
θ_s is the water content of a saturated porous medium
n is a water retention curve fitting parameter (well tabulated empirical parameter)
D_z is the vertical hydrodynamic dispersion coefficient
$\mathcal{D}_e$ is the effective molecular diffusion coefficient
α_z is the vertical dispersivity
τ is the tortuosity coefficient
k_b is Boltzmann's constant
μ is the viscosity of water
r_v is the radius of the virus
λ is the inactivation rate coefficient of viruses in the liquid phase
λ^* is the inactivation rate coefficient of viruses sorbed at the liquid-solid interface
ρ is the bulk density of the soil
K_d is the equilibrium distribution coefficient
k is the liquid to liquid-solid interface mass transfer rate
κ is the liquid to liquid-solid interface mass transfer coefficient
r_p is the average radius of soil particles
k° is the liquid to air-liquid interface mass transfer rate
κ° is the liquid to air-liquid interface mass transfer coefficient
ρ_w is the density of water
g is the acceleration due to gravity
σ is the surface tension of water
α is a water retention curve fitting parameter.
L is the thickness of the barrier.
T is temperature

APPENDIX E

STEADY-STATE SOLUTION TO ON-SITE KINETIC MODEL FOR REMOVAL OF VIRUSES IN SATURATED SOIL

$$\log_{10}\left(\frac{C}{C_0}\right) = \frac{x}{2.3}\frac{\left(1-\sqrt{1+4\alpha_L \lambda/v}\right)}{2\alpha_L}$$

Where $\lambda = \mu_1 + \frac{k_{att1}}{1+k_{\det 1}/\mu_{s1}} + \frac{k_{att2}}{1+k_{\det 2}/\mu_{s2}}$

k_{att} and $k_{\det}$ are attachment and detachment rate coefficients

μ_1 and μ_s are inactivation rate coefficients of free and attached viruses

Subscripts 1 and 2 refer to the 2 different kinetic sites.

λ is equivalent to an overall removal rate coefficient that includes inactivation of free viruses and removal of viruses due to interaction with the 2 kinetic sites. Interaction is the combination of attachment, detachment, and inactivation of attached viruses.

APPENDIX F

EQUATIONS FOR AREA SOURCE MODEL TO ESTIMATE CONCENTRATION OF VIRUSES

$$X(x > x_0, y) = \frac{Q}{\sqrt{2\pi}\bar{u}\sigma_z\{x\}y_0} \times \{\text{Vertical Term}\} \times \{\text{Lateral Term}\} \times \{\text{Decay Term}\}$$

$$\{\text{Vertical Term}\} = 1 + 2\sum_{n=1}^{\infty}\left\{\exp\left[-0.5\left(\frac{2nH_m}{\sigma_z\{x\}}\right)\right]\right\}$$

$$\{\text{Lateral Term}\} = erf\left[\frac{y_0/2 + y}{\sqrt{2}\sigma_y\{x\}}\right] + erf\left[\frac{y_0/2 - y}{\sqrt{2}\sigma_y\{x\}}\right]$$

$$\{\text{Decay Term}\} = \exp(-kx/\bar{u})$$

$$\sigma_z\{x\} = \sigma_{E'}(x + x_0/2) + \sigma_{z0}$$

$$\sigma_y\{x\} = \sigma_{A'}(x + x_0/2)$$

x downwind distance
y crosswind distance from the centerline of the area source
x_0 along wind dimension of the area source
y_0 crosswind dimension of the area source
Q rate of release from the source
$\bar{u}$ mean wind speed
H_m estimated depth of the atmospheric mixing layer
σ_{z0} vertical source dimension
$\sigma_{E'}$ standard deviation of the wind elevation angle in radians
$\sigma_{A'}$ standard deviation of the azimuth wind angle in radians
k inactivation constant

APPENDIX G

STABILITY CATEGORIES REQUIRED FOR AREA SOURCE MODEL

Surface wind speed (m/s)	**Day** Incoming Solar radiation			**Night** Cloud Cover	
	Strong	Moderate	Slight	Thinly Overcast	Mostly Cloudy
< 2	A	A-B	B	-	-
2-3	A-B	B	C	E	F
3-5	B	B-C	C	D	E
5-6	C	C-D	D	D	D
>6	C	D	D	D	D

APPENDIX H

RISK DISTRIBUTIONS FOR GROUNDWATER SCENARIOS

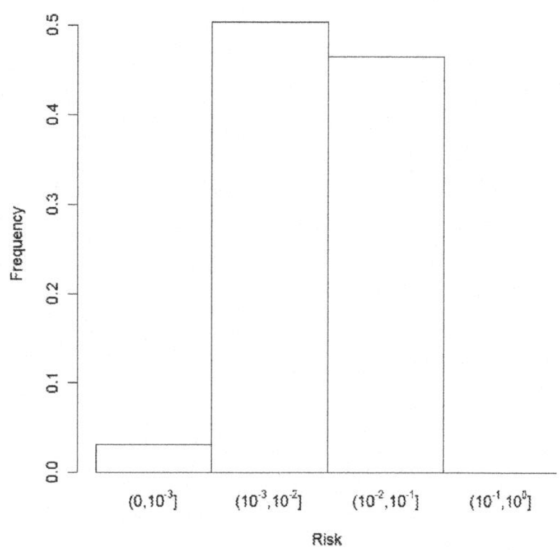

Figure H-1. Risk Distribution for Groundwater Exposure When the Specific Pathway is Non-porous Media ($\mu = 2 \times 10^{-2}$, $\sigma = 2 \times 10^{-2}$).

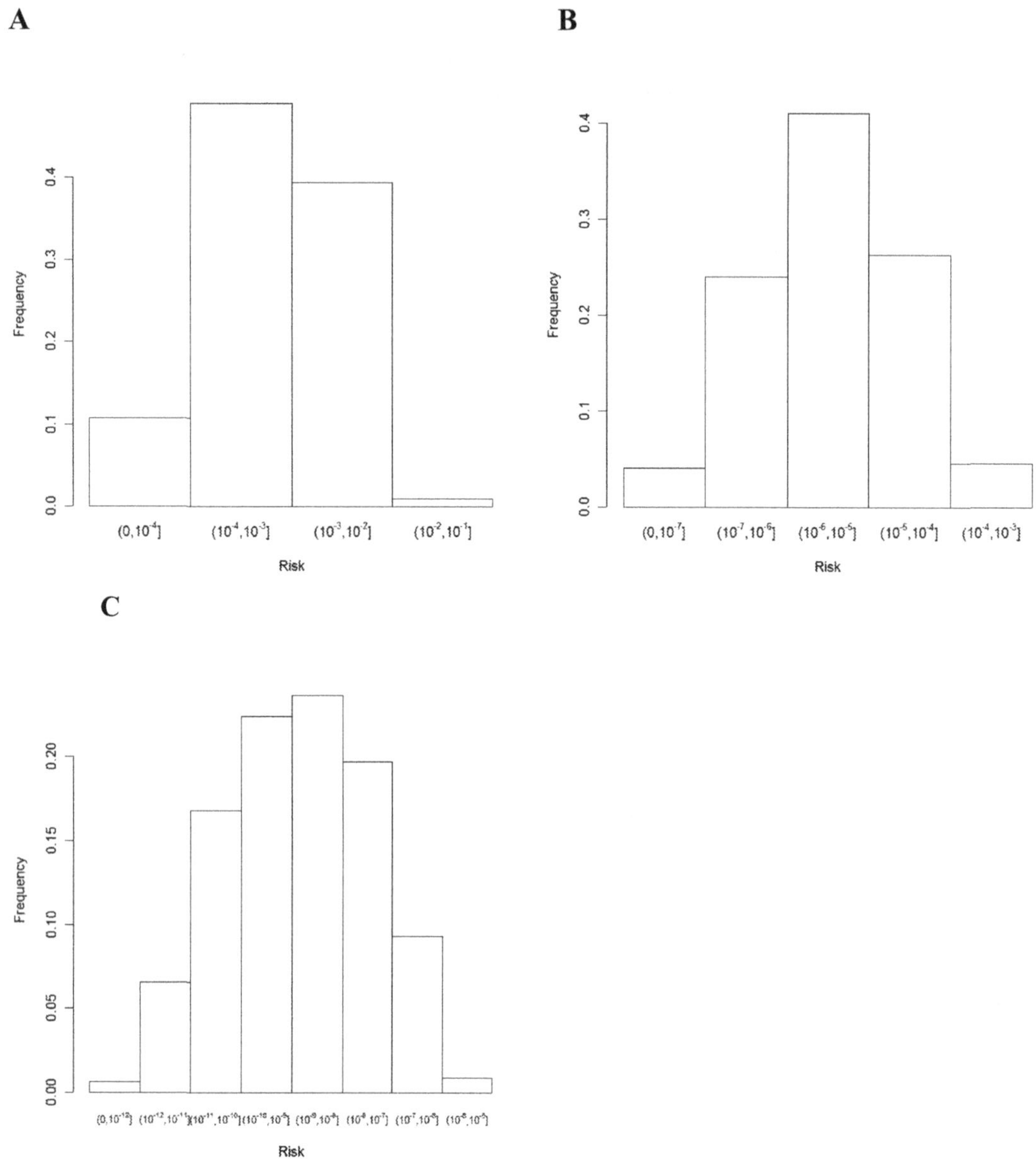

Figure H-2. Risk Distribution For model 4. Non-porous media preceeding, A) 5 meters of saturated Soil: $\mu = 2 \times 10^{-3}$, $\sigma = 2 \times 10^{-3}$, B) 15 meters of saturated Soil: $\mu = 2 \times 10^{-5}$, $\sigma = 4 \times 10^{-5}$, C) 30 meters of saturated soil: $\mu = 5 \times 10^{-8}$, $\sigma = 2 \times 10^{-7}$.

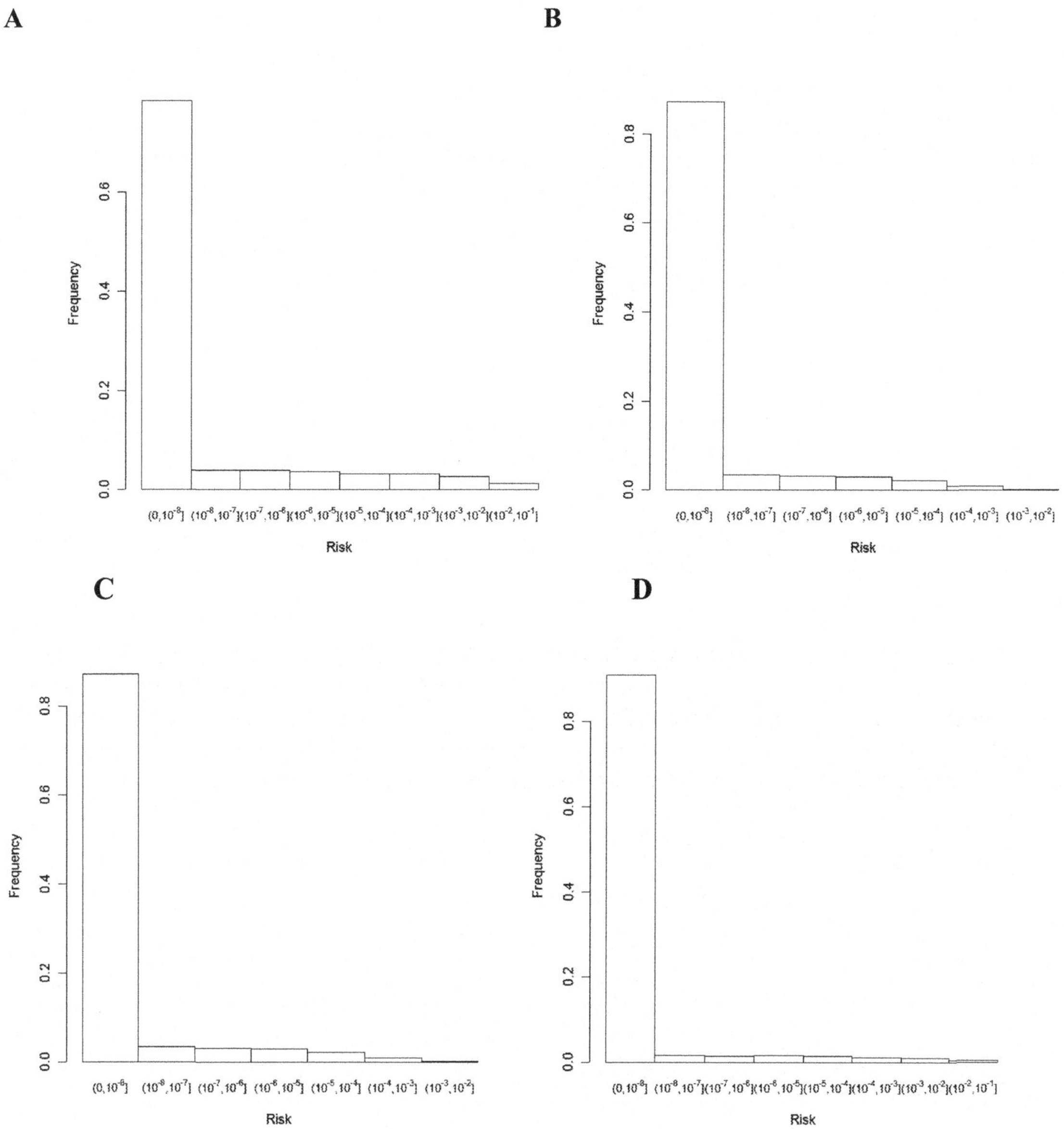
A
B
C
D
Frequency
Risk
0.0
0.2
0.4
0.6
0.8

E **F**

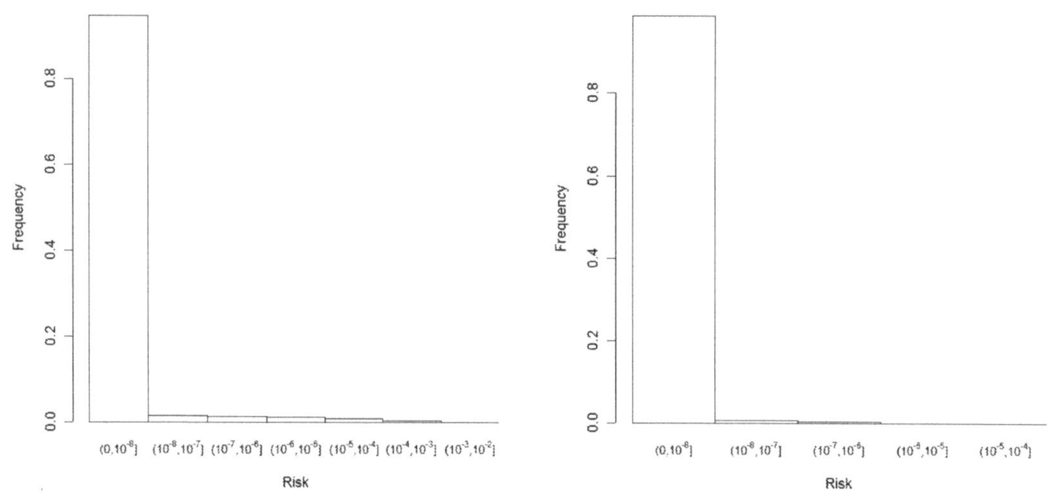

Figure H-3. Risk distribution For Model 4. Unsaturated soil followed by saturated soil, A) Unsaturated Soil = 0.25 meters, Saturated Soil = 5 meters: $\mu = 5 \times 10^{-4}$, $\sigma = 4 \times 10^{-3}$, B) Unsaturated Soil = 0.25 meters, Saturated Soil = 15 meters: $\mu = 6 \times 10^{-6}$, $\sigma = 7 \times 10^{-5}$, C) Unsaturated Soil = 0.25 meters, Saturated Soil = 30 meters: $\mu = 2 \times 10^{-8}$, $\sigma = 3 \times 10^{-7}$, D) Unsaturated Soil = 0.5 meters, Saturated Soil = 5 meters: $\mu = 2 \times 10^{-4}$, $\sigma = 2 \times 10^{-3}$, E) Unsaturated Soil = 0.5 meters, Saturated Soil = 15 meters: $\mu = 2 \times 10^{-6}$, $\sigma = 4 \times 10^{-5}$, F) Unsaturated Soil = 0.5 meters, Saturated Soil = 30 meters: $\mu = 6 \times 10^{-9}$, $\sigma = 2 \times 10^{-7}$.

APPENDIX I

RISK DISTRIBUTIONS FOR AEROSOL SCENARIOS

A

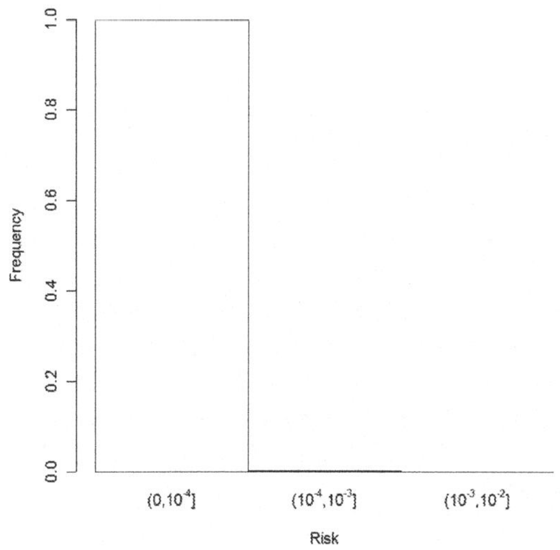

B

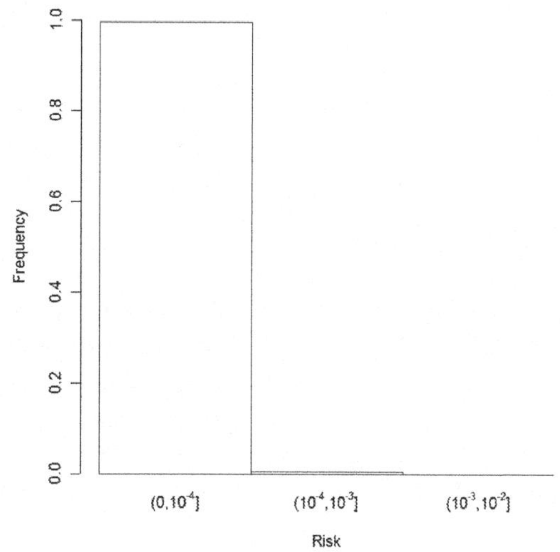

C

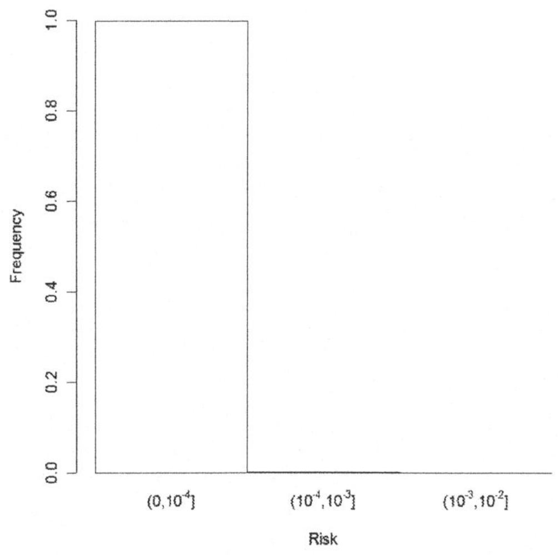

D

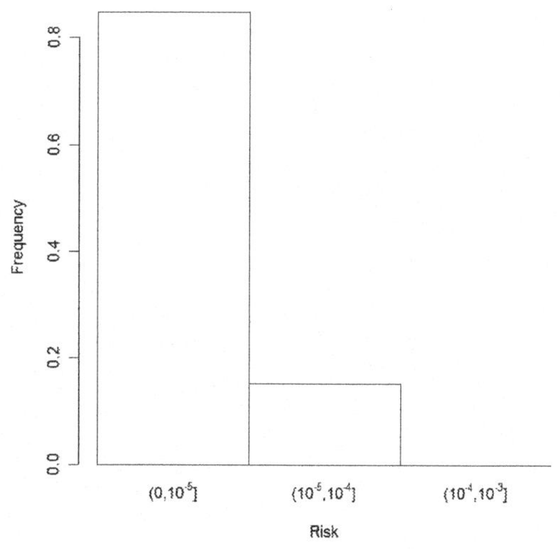

E

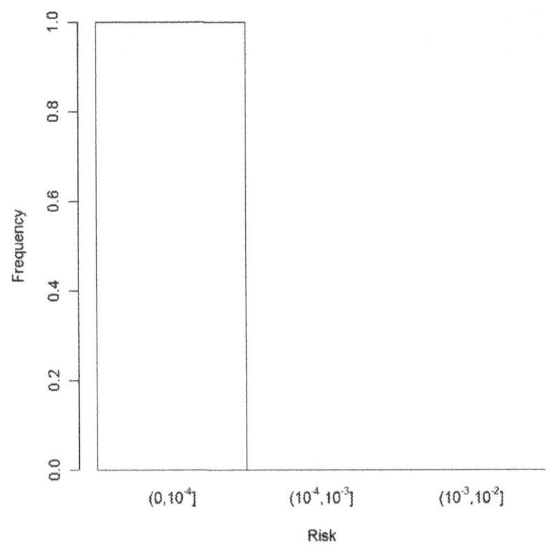

F

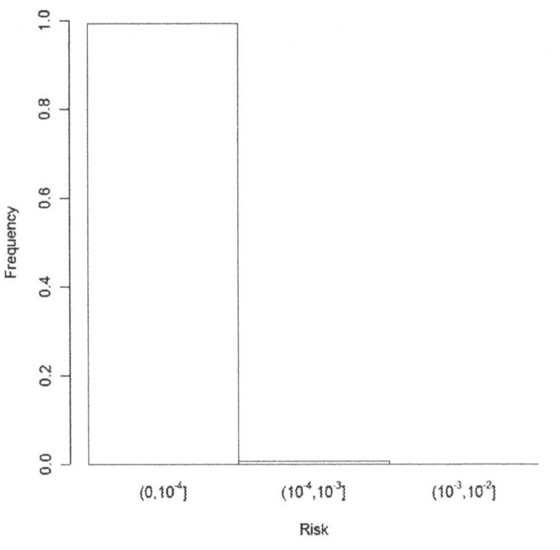

Figure I-1. Risk Distribution For Aerosol Model 4. A) Downwind Distance = 30 meters, Wind Speed = 2 meters per second: mean = 7×10^{-5}, $\sigma = 9 \times 10^{-6}$, B) Downwind Distance = 30 meters, Wind Speed = 5 meters per second: mean = 7×10^{-5}, $\sigma = 9 \times 10^{-6}$, C) Downwind Distance = 30 meters, Wind Speed = 10 meters per second: mean = 7×10^{-5}, $\sigma = 8 \times 10^{-6}$, D) Downwind Distance = 250 meters, Wind Speed = 2 meters per second: mean = 6×10^{-6}, $\sigma = 4 \times 10^{-6}$, E) Downwind Distance = 250 meters, Wind Speed = 5 meters per second: mean = 3×10^{-5}, $\sigma = 8 \times 10^{-6}$, F) Downwind Distance = 250 meters, Wind Speed = 10 meters per second: mean = 5×10^{-5}, $\sigma = 2 \times 10^{-5}$.

REFERENCES

Ahmed, A.U. and D.L. Sorensen. 1995. Kinetics of Pathogen Destruction During Storage of Dewatered Biosolids. *Water Environment Research* 67: 143-150.

Allwood, P.B. 1993. Survival of F-Specific RNA Coliphage, Feline Calicivirus, and Escherichia coli in Water: A comparative Study. *Applied and Environmental Microbiology* 69: 5707-5710.

ASTM. 2005. *Annual Book of ASTM Standards, Section 11, Volume 11.01. Designation: D 4994-89, Standard Practice for Recovery of Viruses from Wastewater Sludges*. American Society for Testing and Materials, Philadelphia.

Bagdasaryan, G.A. 1964. Survival of Viruses of the Enterovirus Group in Soil and on Vegetables. *Journal of Hygiene, Epidemiology, Microbiology and Immunology, 1964, Vol. 8, pp. 497-505.*

Bhattacharjee, S., J.N. Ryan, and M. Elimelech. 2002. Virus transport in physically and geochemically heterogeneous subsurface porous media. *Journal of Contaminant Hydrology* 57: 161-187.

Binder, S., D. Sokal, and D. Maughan. 1986a. Estimating Soil Ingestion: The Use of Tracer Elements in Estimating the Amount of Soil Ingested by Young Children. *Archives of Environmental Health* 41: 341-345.

—. 1986b. Estimating Soil Ingestion: The Use of Tracer Elements in Estimating the Amount of Soil Ingested by Young Children. *Archives of Environmental Health, Vol. 41, November/December 1986, pp. 341-345.*

Bitton, G., O.C. Pancorbo, and S.R. Farrah. 1984. Virus Transport and Survival After Land Application of Sewage Sludge. *Applied and Environmental Microbiology* 47: 905-909.

Brooks, J., B. Tanner, C. Gerba, C. Haas, and I. Pepper. 2005. Estimation of bioaerosol risk of infection to residents adjacent to a land applied biosolids site using an empirically derived transport model. *JOURNAL OF APPLIED MICROBIOLOGY* 98: 397-405.

Brooks, J.P., C.P. Gerba, and I.L. Pepper. 2004. Biological Aerosol Emission, Fate, and Transport from Municipal and Animal Wastes. *Journal of Residuals Science and Technology* 1: 15-28.

Burge, W.D. and N.K. Enkiri. 1978. Virus Adsorption by Five Soils. *Journal of Environmental Quality, 1978, Vol. 7, No. 1, pp. 73-76.*

Calabrese, E.J. 1991. A Gude to Interpreting Soil Ingestion Studies II. Qualitative and Quantitative Evidence of Soil Ingestion. *Regul. Toxicol. Pharmacol, June 1991, Vol. 13, No. 3, pp. 278-292.*

Calabrese, E.J. and E.J. Stanek. 1995. Resolving Intertracer Inconsistencies in Soil Ingestion Estimation. *Environmental Health Perspectives* 103: 454-457.

Calabrese, E.J., E.J. Stanek, R.C. James, and S.M. Roberts. 1997a. Soil Ingestion - a Concern for Acute Toxicity in Children. *Environmental Health Perspectives* 105: 1354-1358.

Calabrese, E.J., E.J. Stanek, P. Pekow, and R.M. Barnes. 1997b. Soil Ingestion Estimates for Children Residing on a Superfund Site. *Ecotoxicity and Environmental Safety* 36: 258-268.

—. 1997c. Soil Ingestion Estimates for Children Residing on a Superfund Site. *Ecotoxicology & Environmental Safety* 36: 258-268.

Chattopadhyay, S., and R.W. Puls. 1999a. Adsorption of Bacteriophages on Clay Minerals. *Environmental Science and Technology, 33: 3609-3614.*

—. 1999b. Forces dictating colloidal interactions between viruses and soil. *Chemospere, 41: 1279-1286.*

—. 2002. Effects of Surfactants on the Survival and Adsorption of Viruses. *Environmental Science and Technology, 36 (19): 4017-4024.*

Chetochine, A.S., M.L. Brusseau, C. Gerba, and I. Pepper. 2006. Leaching of Phage from Class B Biosolids and Potential Transport through Soil *Applied and Environmental Microbiology* 72: 665-671.

Chu, Y., Y. Jin, M. Flury, and M.V. Yates. 2001. Mechanisms of virus removal during transport in unsaturated porous media. *Water Resources Research* 37: 253-263.

Clausing, P., B. Brunekreef, and J.H. van Wijnen. 1987a. A method for estimating soil ingestion by children. *Int Arch Occup Environ Health* 59: 73-82.

—. 1987b. A method for estimating soil ingestion by children. *Int Arch Occup Environ Health (1987) 59:73-82.*

Colford, J., D. Eisenberg, J.N.S. Eisenberg, J. Scott, and J.A. Soller. 2003. A Dynamic Model to Assess Microbial Health Risks Associated with Beneficial Uses of Biosolids - Phase 1. *Water Environment Research Foundation Report 98-REM-1.*

Deborde, D.C., W.W. Woessner, Q.T. Kiley, and P. Ball. 1999. Rapid transport of viruses in a floodplain aquifer. *Water Research* 33: 2229-2238.

Dowd, S.E., C.P. Gerba, I.L. Pepper, and S.D. Pillai. 2000. Bioaerosol transport modeling and risk assessment in relation to biosolid placement. *Journal of Environmental Quality* 29.

Dowd, S.E. and S.D. Pillai. 1997. Survival and Transport of Selected Bacterial Pathogens and Indicator Viruses under Sandy Aquifer Conditions. *Journal of Environmental Science & Health, Part A: Environmental Science & Engineering & Toxic & Hazardous Substance Control* 32: 2245-2258.

Eisenberg, J.N., M.A. Brookhart, G. Rice, M. Brown, and J.M. Colford, Jr. 2002. Disease transmission models for public health decision making: analysis of epidemic and endemic conditions caused by waterborne pathogens. *Environ Health Perspect* 110: 783-90.

Eisenberg, J.N., E.Y.W. Seto, A. Olivieri, and R.C. Spear. 1996. Quantifying water pathogen risk in an epidemiological framework. *Risk Analysis* 16: 549-563.

Eisenberg, J.N.S., J.A. Soller, J. Scott, D. Eisenberg, and J. Colford. 2004. A Dynamic Model to Assess Microbial Health Risks Associated with Beneficial Uses of Biosolids. *Risk Analysis* 24: 221-236.

Faulkner, B.R., W.G. Lyon, F.A. Khan, and S. Chattopadhyay. 2002. Predicting Attenuation ofViruses During Percolation in Soils. *EPA/600/R-02/051a.*

—. 2003. Modeling leaching of viruses by the Monte Carlo method. *Water Research* 37: 4719-4729.

Finley, B., D. Proctor, P. Scott, N. Harringtoin, D. Paustenbach, and P. Price. 1994. Recommended Distributions for Exposure Factors Frequently Used in Health Risk Assessment. *Risk Analysis* 14: 533-551.

Gerba, C.P., I.L. Pepper, and L.F. Whitehead. 2002. A risk assessment of emerging pathogens of concern in the land application of biosolids [Review]. *Water Science and Technology* 46: 225-230.

Gerba, C.P., C. Wallis, and J.L. Melnick. 1975. Fate of Wastewater Bacteria and Viruses in Soil. *Journal of the Irrigation and Drainage Division, September 1975, pp. 157-174.*

Goddard. 1981. Recovery of Indigenous Enteroviruses from Raw and Digested Sewage Sludges. *Applied and Environmental Microbiology, December 1981, Vol. 42, No. 6, pp. 1023-1028.*

Goyal, S.M., and C.P. Gerba. 1979. Comparative Adsorption of Human Enteroviruses, Simian Rotovirus, and Selected Bacteriophages to Soils. *Applied and Environmental Microbiology, August 1979, Vol. 38, No. 2, pp. 241-247.*

Haas, C.N. 1983. Estimation of Risk Due to Low Doses of Microorganisms: A Comparison of Alternative Methodologies. *American Journal of Epidemiology* 55: 573-582.

Hamparian, V.V., A.C. Ottolenghi, and J.H. Hughes. 1985. Enteroviruses in sludge: multiyear experience with four wastewater treatment plants. *Applied and Environmental Microbiology* 50: 280-286.

Hawley, J.K. 1985a. Assessment of Health Risk from Exposure to Contaminated Soil. *Risk Analysis* 5: 289-302.

—. 1985b. Assessment of Health Risk from Exposure to Contaminated Soil. *Risk Analysis, 1985, Vol. 5, No. 4, pp. 289-302.*

Hurst, C.J. 1989. Fate of Viruses During Wastewater Sludge Treatment Processes. *CRC Critical Reviews in Environmental Control* 18: 317-343.

International Life Sciences Institute, G. Craun, A. Dufour, J. Eisenberg, J. Foran, C. Gauntt, et al. 1996. A conceptual framework to assess the risks of human disease following exposure to pathogens. *Risk Analysis* 16: 841-848.

LaGoy, P.K. 1987a. Estimated Soil Ingestion Rates for Use in Risk Assessment. *Risk Analysis* 7: 355-359.

—. 1987b. Estimated Soil Ingestion Rates for Use in Risk Assessment. *Risk Analysis, Vol. 7, No. 3, 1987.*

Lance, J. and C. Gerba. 1984. Virus Movement in Soil During Saturated and Unsaturated Flow. *Applied and Environmental Microbiology, 47 (2):335-337.*

Matthess, G., A. Pekdeger, and J. Schroefer. 1988. Persistence and transport of bacteria and viruses in groundwater--a conceptual evaluation. *J Cont Hyd* 2: 171-188.

Mawdsley, J.L., R.D. Bardgett, R.J. Merry, B.F. Pain, and M.K. Theodorou. 1995. Pathogens in livestock waste, their potential for movement through soil and environmental pollution [Review]. *Applied Soil Ecology* 2: 1-15.

Meschke, J.S., and M.D. Sobsey. 2003. Comparative reduction of Norwalk virus, poliovirus type 1, F+ RNA coliphage MS2 and Escherichia coli in miniature soil columns. *Water Science and Technology* 47: 85-90.

Nasser, A.M., R. Glozman, and Y. Nitzan. 2002. Contribution of microbial activity to virus reduction in saturated soil. *Water Research* 36: 2589-2595.

NRC. 1983. *Risk assessment in the federal government: managing the process.* National Research Council, National Academy Press, Washington, D.C.

—. 2002. *Biosolids applied to Land: Advancing standards and practices* National Academies Press Washington, D.C.

Parker, D.T., J.C. Spendlove, J.A. Bondurant, and J.H. Smith. 1977. Microbial aerosols from food processing waste spray fields. *Journal Water Pollution Control Federation* 49: 2359-2365.

Pasquill, F. 1962. *Atmospheric diffusion; the dispersion of windborne material from industrial and other sources*. Van Nostrand, London, Princeton, NJ.

Pedley, S., M. Yates, J. Schijven, W.J. J. Howard, and M. Barrett. 2006. Pathogens: Health relevance, transport and attenuation. *Protecting groundwater for health: managing the quality of drinking-water sources* World Health Organization, Geneva.

Pillai, S.D. 2002. Bioaerosols and public health. *Microbiology of Composting*: 595-606.

Pillai, S.D. and S.C. Ricke. 2002. Bioaerosols from municipal and animal wastes: background and contemporary issues [Review]. *Canadian Journal of Microbiology* 48: 681-696.

Reddy, K., R. Khaleel, and M.R. Overcash. 1981. Behavior and Transport of Microbial Pathogens and Indicator Organisms in Soils Treated with Organic Wastes. *J. Environ. Qual. 10:255-265.*

Redman, J., M. Estes, and S. Grant. 2001. Resolving macroscale and microscale heterogeneity in virus filtration. *Colloids and Surfaces A: Physiochem. Eng. Aspects* 191: 57-70.

Regli, S., J.B. Rose, C.N. Haas, and C.P. Gerba. 1991a. Modeling the Risk from Giardia and Viruses in Drinking-Water. *JAWWA* 83: 76-84.

—. 1991b. Modeling the Risk From Giardia and Viruses in Drinking Water. *Journal American Water Works Association* 83: 76-84.

Rosebury, A.M., and D.E. Burmaster. 1992. Lognormal distributions for water intake by children and adults. *Risk Analysis* 12: 99-104.

Sanin, F.D., P.A. Vesilind, and C.J. Martel. 1994. Pathogen Reduction Capabilities of Freeze/Thaw Sludge Conditioning. *Water Research* 28: 2393-2398.

Schijven, J. and S. Hassanizadeh. 2000a. Removal of viruses by soil passage: Overview of modeling, processes, and parameters. *CRITICAL REVIEWS IN ENVIRONMENTAL SCIENCE AND TECHNOLOGY* 30: 49-127.

Schijven, J., S. Hassanizadeh, and R. de Bruin. 2002. Two-site kinetic modeling of bacteriophages transport through columns of saturated dune sand. *Journal Of Contaminant Hydrology* 57: 259-279.

Schijven, J.F., H.A.M. de Bruin, S.M. Hassanizadeh, and A.M.D. Husman. 2003. Bacteriophages and clostridium spores as indicator organisms for removal of pathogens by passage through saturated dune sand. *Water Research* 37: 2186-2194.

Schijven, J.F. and S.M. Hassanizadeh. 2000b. Removal of viruses by soil passage: Overview of modeling, processes, and parameters [Review]. *Critical Reviews in Environmental Science & Technology* 30: 49-127.

Schijven, J.F., G. Medema, A.J. Vogelaar, and S.M. Hassanizadeh. 2000. Removal of microorganisms by deep well injection. *Journal of Contaminant Hydrology* 44: 301-327.

Schijven, J.F. and J. Simunek. 2002. Kinetic modeling of virus transport at the field scale. *Journal of Contaminant Hydrology* 55: 113-135.

Shumway, R.H., R.S. Azari, and M. Kayhanian. 2002. Statistical Approaches to Estimating Mean Water Quality Concentrations with Detection Limits. *Environ. Sci. Technol.* 36: 3345-3353.

Sim, Y. and C.V. Chrysikopoulos. 2000. Virus transport in unsaturated porous media. *Water Resources Research* 36: 173-179.

Simon, S. 1998. Soil Ingestion by humans: a review of history, data, and etiology with application to risk assessment fo radioactivity contaminated soil. *Health Phys, 74 (6): 647-672.*

Soares, A., T. Straub, I. Pepper, and C. Gerba. 1994. Effect of anaerobic digestion on the occurrence of enteroviruses and *Giardia* cysts in sewage sludge. *Journal Environmental Science Health* A29: 1887-1897.

Sobsey, M.D., C.H. Dean, M.E. Knuckles, and R.A. Wagner. 1980. Interactions and Survival of Enteric Viruses in Soil Materials. *Applied and Environmental Microbiology, July 1980, Vol. 40, No. 1, pp. 92-101.*

Soller, J.A., J.N. Eisenberg, and A.W. Olivieri. 1999. Evaluation of Pathogen Risk Assessment Framework. Prepared for the International Life Sciences Institute, Risk Science Institute., Washington, D.C.

Soller, J.A., A.W. Olivieri, J. Crook, R.C. Cooper, G. Tchobanoglous, R.T. Parkin, et al. 2003. Risk-based approach to evaluate the public health benefit of additional wastewater treatment. *Environ Sci Technol* 37: 1882-91.

Sorber, C.A. and B. Moore. 1986a. Survival and Transport of Pathogens in Sludge-Amended Soil, A Critcal Literature Review. *Office of Research and Development, U.S. EPA, Cooperative Agreement No. CR811918-01-0, 1986.*

—. 1986b. Survival and Transport of Pathogens in Sludge-Amended Soil, A Critical Literature Review. Office of Research and Development, U.S. Environmental Protection Agency, Washinton, D.C.

Stanek, E.J., 3rd, and E. J. Calabrese. 2000. Daily soil ingestion estimates for children at a Superfund site. *Risk Analysis* 20: 627-35.

Stanek, E.J. and E.J. Calabrese. 1995. Daily Estimates of Soil Ingestion in Children. *Environmental Health Perspective, March 1995, Vol. 103, No. 3, pp. 276-285.*

Stanek, E.J., E.J. Calabrese, R. Barnes, and P. Pekow. 1997. Soil Ingestion in Adults- Results of a Second Pilot Study. *Ecotoxicity and Environmental Safety, 1997, Vol. 36, pp. 249-257.*

Stanek, E.J., E.J. Calabrese, and M. Zorn. 2001. Biasing factors for simple soil ingestion estimates in mass balance studies of soil ingestion. *Human & Ecological Risk Assessment* 7: 329-355.

Teltsch, B. and E. Katzenelson. 1978. Airborne enteric bacteria and viruses from spray irrigation with wastewater. *Appl. Environ. Microbiol.* 35: 290-296.

Teunis, P.F.M., O.G. van der Heijden, J.W.B. van der Giessen, and A.H. Havelaar. 1996. The dose-response relation in human volunteers for gastro-intestinal pathogens. National Institute of Public Health and the Environment, Bilthoven, The Netherlands.

Thompson, K.M. and D.E. Burmaster. 1991. Parametric Distributions for Soil Ingestion by Children. *Risk Analysis* 11: 339-342.

Tierney, J.T., R. Sullivan, and E.P. Larkin. 1977. Persistence of Poliovirus 1 in Soil and on Vegetables Grown in Soil Previously Flooded with Inoculated Sewage Sludge or Effluent. *Applied and Environmental Microbiology, Jan 1977, Vol. 33, No. 1, pp. 109-113.*

U.S. EPA. 1988. Preliminary Risk Assessment for Viruses in Municipal Sewage Sludge Applied to Land.

—. 1989a. Interim Final Guidance on Soil Ingestion Rates. U.S. Environmental Protection Agency, Washington, D.C.

—. 1989b. Pathogen Risk Assessment for Land Application of Municipal Sludge. Volume I: Methodology and computer model. U.S. Environmental Protection Agency, Washington, D.C.

—. 1991a. Pathogen risk assessment for land application of municipal sludge. Office of Health and Environmental Assessment, United States Environmental Protection Agency, Washington, D.C.

—. 1991b. Preliminary risk assessment for bacteria in municipal sewage sludge applied to land. Office of Health and Environmental Assessment, United States Environmental Protection Agency, Washington, D.C.

—. 1992a. Preliminary risk assessment for viruses in municipal sewage sludge applied to land. Office of Health and Environmental Assessment, United States Environmental Protection Agency, Washington, D.C.

—. 1992b. Technical Support Document for Reduction of Pathogens and Vector Attraction in Sewage Sludge. Office of Health and Environmental Assessment, United States Environmental Protection Agency, Washington, D.C.

—. 1993a. A Plain English Guide to the EPA Part 503 Biosolids Rule. United States Environmental Protection Agency, Washington, D.C.

—. 1993b. Standards for the Use of Disposal of Sewage Sludge, Final Rules. U.S. Environmental Protection Agency, Washington, D.C.

—. 1995. A Guide to the Biosolids Risk Assessment for the EPA Part 503 Rule. U.S. Environmental Protection Agency, Washington, D.C.

—. 2000. Estimated Per Capita Water Ingestion in the United States: Based on Data Collected by the United States Department of Agriculture's 1994-96 Continuing Survey of Food Intakes by Individuals. United States Environmental Protection Agency, Washington, D.C.

—. 2006. Fact Sheet - Long Term 2 Enhanced Surface Water Treatment Rule. U.S. Environmental Protection Agency.

Ward, R.L., and C.S. Ashley. 1977. Inactivation of Enteric Viruses in Wastewater Sludge Through Dewatering by Evaporation. *Applied and Environmental Microbiology, Nov. 1977, Vol. 34, No. 5, pp. 564-570.*

Yates, M., C. Gerba, and L. Kelley. 1985. Virus Peristence in Groundwater. *Applied and Environmental Microbiology* 49: 778-781.

Zhuang, J. and Y. Jin. 2003a. Virus retention and transport as influenced by different forms of soil organic matter. *Journal of Environmental Quality* 32: 816-823.

—. 2003b. Virus retention and transport through Al-oxide coated sand columns: effects of ionic strength and composition. *Journal of Contaminant Hydrology* 60: 193-209.

WERF SUBSCRIBERS

WASTEWATER UTILITY

Alabama
Montgomery Water Works & Sanitary Sewer Board

Alaska
Anchorage Water & Wastewater Utility

Arizona
Glendale, City of, Utilities Department
Mesa, City of
Peoria, City of
Phoenix Water Services Department
Pima County Wastewater Management
Safford, City of

Arkansas
Little Rock Wastewater Utility

California
Calaveras County Water District
Central Contra Costa Sanitary District
Corona, City of
Crestline Sanitation District
Delta Diablo Sanitation District
Dublin San Ramon Services District
East Bay Dischargers Authority
East Bay Municipal Utility District
Eastern Municipal Water District
El Dorado Irrigation District
Fairfield-Suisun Sewer District
Fresno Department of Public Utilities
Inland Empire Utilities Agency
Irvine Ranch Water District
Las Virgenes Municipal Water District
Livermore, City of
Los Angeles, City of
Los Angeles County, Sanitation Districts of
Napa Sanitation District
Orange County Sanitation District
Palo Alto, City of
Riverside, City of
Sacramento Regional County Sanitation District
San Diego Metropolitan Wastewater Department, City of
San Francisco, City & County of
San Jose, City of
Santa Barbara, City of
Santa Cruz, City of
Santa Rosa, City of
South Bayside System Authority
South Coast Water District
South Orange County Wastewater Authority
Stege Sanitary District
Sunnyvale, City of
Union Sanitary District
West Valley Sanitation District

Colorado
Aurora, City of
Boulder, City of
Greeley, City of
Littleton/Englewood Water Pollution Control Plant
Metro Wastewater Reclamation District, Denver

Connecticut
Greater New Haven WPCA

District of Columbia
District of Columbia Water & Sewer Authority

Florida
Broward, County of
Fort Lauderdale, City of
Miami-Dade Water & Sewer Authority
Orange County Utilities Department
Reedy Creek Improvement District
Seminole County Environmental Services
St. Petersburg, City of
Stuart Public Utilities
Tallahassee, City of
Tampa, City of
Toho Water Authority
West Palm Beach, City of

Georgia
Atlanta Department of Watershed Management
Augusta, City of
Clayton County Water Authority
Cobb County Water System
Columbus Water Works
Fulton County
Gwinnett County Department of Public Utilities
Savannah, City of

Hawaii
Honolulu, City & County of

Idaho
Boise, City of

Illinois
American Bottoms Wastewater Treatment Plant
Greater Peoria Sanitary District
Kankakee River Metropolitan Agency
Metropolitan Water Reclamation District of Greater Chicago
Wheaton Sanitary District

Iowa
Ames, City of
Cedar Rapids Wastewater Facility
Des Moines, City of
Iowa City

Kansas
Johnson County Unified Wastewater Districts
Unified Government of Wyandotte County/ Kansas City, City of

Kentucky
Louisville & Jefferson County Metropolitan Sewer District

Louisiana
Sewerage & Water Board of New Orleans

Maine
Bangor, City of
Portland Water District

Maryland
Anne Arundel County Bureau of Utility Operations
Howard County Department of Public Works
Washington Suburban Sanitary Commission

Massachusetts
Boston Water & Sewer Commission
Upper Blackstone Water Pollution Abatement District

Michigan
Ann Arbor, City of
Detroit, City of
Holland Board of Public Works
Saginaw, City of
Wayne County Department of Environment
Wyoming, City of

Minnesota
Rochester, City of
Western Lake Superior Sanitary District

Missouri
Independence, City of
Kansas City Missouri Water Services Department
Little Blue Valley Sewer District
Metropolitan St. Louis Sewer District

Nebraska
Lincoln Wastewater System

Nevada
Henderson, City of
Reno, City of

New Jersey
Bergen County Utilities Authority
Ocean, County of
Passaic Valley Sewerage Commissioners

New York
New York City Department of Environmental Protection

North Carolina
Charlotte/Mecklenburg Utilities
Durham, City of
Metropolitan Sewerage District of Buncombe County
Orange Water & Sewer Authority

Ohio
Akron, City of
Butler County Department of Environmental Services
Columbus, City of
Metropolitan Sewer District of Greater Cincinnati
Northeast Ohio Regional Sewer District
Summit, County of

Oklahoma
Oklahoma City Water & Wastewater Utility Department
Tulsa, City of

Oregon
Clean Water Services
Eugene/Springfield Water Pollution Control
Gresham, City of
Water Environment Services

Pennsylvania
Philadelphia, City of
University Area Joint Authority

South Carolina
Charleston Commissioners of Public Works
Mount Pleasant Waterworks & Sewer Commission
Spartanburg Sanitary Sewer District

Tennessee
Cleveland, City of
Knoxville Utilities Board
Murfreesboro Water & Sewer Department
Nashville Metro Water Services

Texas
Austin, City of
Dallas Water Utilities
Denton, City of
El Paso Water Utilities
Fort Worth, City of
Gulf Coast Waste Disposal Authority
Houston, City of
San Antonio Water System
Trinity River Authority

Utah
Salt Lake City Corporation

Virginia
Alexandria Sanitation Authority

Arlington, County of
Fairfax County
Hampton Roads Sanitation District
Henrico, County of
Hopewell Regional Wastewater Treatment Facility
Loudoun County Sanitation Authority
Lynchburg Regional WWTP
Prince William County Service Authority
Richmond, City of
Rivanna Water & Sewer Authority

Washington

Everett, City of
King County Department of Natural Resources
Seattle Public Utilities
Sunnyside, Port of
Yakima, City of

Wisconsin

Green Bay Metro Sewerage District
Kenosha Water Utility
Madison Metropolitan Sewerage District
Milwaukee Metropolitan Sewerage District
Racine, City of
Sheboygan Regional Wastewater Treatment
Wausau Water Works

Australia

South Australian Water Corporation
Sydney Water Corporation
Water Corporation of Western Australia

Canada

Greater Vancouver Regional District
Regina, City of, Saskatchewan
Toronto, City of, Ontario
Winnipeg, City of, Manitoba

New Zealand

Watercare Services Limited

United Kingdom

Yorkshire Water Services Limited

STORMWATER UTILITY

California

Los Angeles, City of, Department of Public Works
Monterey, City of
Sacramento, County of
San Francisco, City & County of
Santa Rosa, City of
Sunnyvale, City of

Colorado

Aurora, City of
Boulder, City of

Georgia

Griffin, City of

Iowa

Cedar Rapids Wastewater Facility
Des Moines, City of

Kansas

Overland Park, City of

Maine

Portland Water District

Minnesota

Western Lake Superior Sanitary District

North Carolina

Charlotte, City of, Stormwater Services

Pennsylvania

Philadelphia, City of

Tennessee

Chattanooga Stormwater Management

Texas

Harris County Flood Control District, Texas

Washington

Bellevue Utilities Department
Seattle Public Utilities

STATE

Arkansas Department of Environmental Quality
Connecticut Department of Environmental Protection
Fresno Metropolitan Flood Control District
Kansas Department of Health & Environment
Kentucky Department of Environmental Protection
Ohio River Valley Sanitation Commission
Urban Drainage & Flood Control District, CO

CORPORATE

ADS Environmental Services
Alan Plummer & Associates
Alden Research Laboratory Inc.
Alpine Technology Inc.
Aqua-Aerobic Systems Inc.
Aquateam–Norwegian Water Technology Centre A/S
ARCADIS
Associated Engineering
Black & Veatch
Boyle Engineering Corporation
Brown & Caldwell
Burgess & Niple, Ltd.
Burns & McDonnell
CABE Associates Inc.
The Cadmus Group
Camp Dresser & McKee Inc.
Carollo Engineers Inc.
Carpenter Environmental Associates Inc.
CDS Technologies Inc.
CET Engineering Services
Chemtrac Systems Inc.
CH2M HILL
Conestoga-Rovers & Associates
CONTECH Stormwater Solutions
D&B/Guarino Engineers, LLC
Damon S. Williams Associates, LLC
Earth Tech Inc.
Ecovation
EMA Inc.
Environ/The ADVENT Group, Inc.
Fay, Spofford, & Thorndike Inc.
Freese & Nichols Inc.
ftn Associates Inc.
Fuss & O'Neill Inc.
Gannett Fleming Inc.
Geosyntec Consultants
GHD
Golder Associates Ltd.
Greeley and Hansen LLC
Hazen & Sawyer, P.C.
HDR Engineering Inc.
HNTB Corporation
Hydromantis Inc.
HydroQual Inc.
Infilco Degremont Inc.
Jacobson Helgoth Consultants Inc.
Jason Consultants LLC Inc.
Jordan, Jones, & Goulding Inc.
KCI Technologies Inc.
Kelly & Weaver, P.C.
Kennedy/Jenks Consultants
KMK Consultants
Komline Sanderson Engineering Corporation
Limno-Tech Inc.
Lombardo Associates Inc.
The Low Impact Development Center Inc.
Malcolm Pirnie Inc.
Material Matters
McKim & Creed
MEC Analytical Systems, Inc.
Metcalf & Eddy Inc.
Monteco Corporation
MPR Engineering Corporation, Inc.
MWH
O'Brien & Gere Engineers Inc.
Odor & Corrosion Technology Consultants Inc.
Oscar Larson & Associates
PA Government Services Inc.
Parametrix Inc.
Parsons
Post, Buckley, Schuh & Jernigan
RMC Water & Environment
R.M. Towill Corporation
Ross & Associates Ltd.
Rothberg, Tamburini & Windsor, Inc.
SAIC
Siemens Water Technologies
Stantec Consulting Inc.
Stearns & Wheler, LLC
Stone Environmental Inc.
Stratus Consulting Inc.
Synagro Technologies Inc.
Tetra Tech Inc.
Trojan Technologies Inc.
Trussell Technologies, Inc.
URS Corporation
Wade-Trim Inc.
Westin Engineering Inc.
Weston Solutions Inc.
Woodard & Curran
Zenon Environmental Inc.
Zoeller Pump Company

INDUSTRY

American Electric Power
American Water
ChevronTexaco Energy Research & Technology Company
The Coca-Cola Company
Dow Chemical Company
DuPont Company
Eastman Chemical Company
Eastman Kodak Company
Eli Lilly & Company
Merck & Company Inc.
Premier Chemicals LLC
Procter & Gamble Company
RWE Thames Water Plc
Severn Trent Services Inc.
Suez Environnment
United Water Services LLC

Note: List as of 9/1/06

IWA Publishing's authorised EU representative for General Product Safety Regulations is Diane D'Arras, 15 rue Duret, 75116 Paris, France, e-mail: safety@iwap.co.uk.

Printed and bound by CPI Group (UK) Ltd, Croydon, CR0 4YY
17/09/2026
02218811-0010